Full-Duplex Integrated Sensing and Communication Systems

Changhao Du · Zhongshan Zhang · Jianping An ·
Xinyuan Zhang · Hongru Zhang · Jie Yang

Full-Duplex Integrated Sensing and Communication Systems

Principles, Key Technologies, and Receiver Design

Changhao Du
School of Cyberspace Science and Technology
Beijing Institute of Technology
Beijing, China

Jianping An
Beijing Institute of Technology
Beijing, China

Hongru Zhang
School of Information and Electronics
Beijing Institute of Technology
Beijing, China

Zhongshan Zhang
School of Cyberspace Science and Technology
Beijing Institute of Technology
Beijing, China

Xinyuan Zhang
School of Information and Electronics
Beijing Institute of Technology
Beijing, China

Jie Yang
School of Cyberspace Science and Technology
Beijing Institute of Technology
Beijing, China

ISBN 978-981-92-0469-4 ISBN 978-981-92-0470-0 (eBook)
https://doi.org/10.1007/978-981-92-0470-0

This Springer imprint is published by the registered company Springer Nature Singapore Pte Ltd.
The registered company address is: 152 Beach Road, #21-01/04 Gateway East, Singapore 189721, Singapore

Preface

With the large-scale commercial deployment of the fifth-generation (5G) mobile communication technology and the ongoing development of the sixth-generation (6G), wireless communication systems are evolving towards higher spectral efficiency, lower latency, and greater intelligence. In this book, Integrated Sensing and Communication (ISAC) technology, which enables the deep integration of communication and sensing functionalities on a single platform, has emerged as a key enabler for future wireless networks. However, traditional ISAC systems, such as Time-Division Duplexing (TDD)-based systems, face challenges such as sensing blind spots, limited communication capacity, and complex temporal scheduling, which severely restrict their application in highly dynamic scenarios.

The introduction of Full-Duplex (FD) technology brings a revolutionary breakthrough for ISAC systems. By enabling simultaneous transmission and reception in the same frequency band, FD-ISAC not only eliminates the sensing blind spots inherent in traditional TDD schemes but also significantly enhances communication throughput, realizing the true vision of "Communication enables Sensing, and Sensing enhances Communication." This book systematically elaborates on the theoretical foundations, key algorithms, hardware implementation, and future trends of FD-ISAC, aiming to provide a professional reference for researchers, engineers, and senior postgraduate students that combines theoretical depth with practical guidance.

The book is structured into three distinct parts, each building upon the last in a logical progression:

Part I: Theoretical Foundations of FD-ISAC

This part begins with the basic models of ISAC systems and delves into core technologies such as waveform design, channel modeling, and full-duplex self-interference cancellation: Chapter 1 introduces the background, technical advantages of FD-ISAC, and its applications in typical scenarios like intelligent transportation, the low-altitude economy, and smart homes. Chapter 2 systematically outlines three main paths for ISAC waveform design, including radar-centric, communication-centric, and joint design, while establishing comprehensive communication and sensing channel models. Chapter 3 focuses on the FD self-interference (SI) problem, providing a detailed analysis of linear and nonlinear modeling for SI signals, and systematically introduces tri-domain (antenna, radio frequency, and digital) SI suppression techniques.

Part II: Algorithm and Implementation of FD-ISAC

This part focuses on the signal processing flow, hardware platform design, and on-chip integration solutions for FD-ISAC systems: Chapter 4 proposes a received signal processing flow tailored for FD-ISAC, including algorithms for separating target echoes from SI signals, and introduces various radar signal processing methods. Chapter 5 provides a comprehensive review of the hardware implementation paths for FD SI Cancellation (SIC), covering key technologies such as antenna isolation, RF cancellation, and digital filtering, alongside performance measurements from several prototype systems. Chapter 6 concentrates on the design of on-chip RF cancellers, systematically comparing the advantages and disadvantages of four mainstream architectures, including Frequency-Domain Equalization (FDE), Time-Domain Equalization (TDE), Hilbert-Transform Equalization (HTE), and MIMO cancellation.

Part III: Current Challenges and Emerging Frontiers in FD-ISAC

This part summarizes the core challenges facing FD-ISAC in its current stage of development and explores its convergence with cutting-edge technologies such as Terahertz (THz), Reconfigurable Intelligent Surfaces (RIS), and Artificial Intelligence (AI): Chapter 7 analyzes the technical bottlenecks of FD-ISAC from three dimensions, including model design, signal processing, and on-chip implementation. Chapter 8 discusses the potential of FD-ISAC in emerging fields like Terahertz communications, RIS-assisted systems, AI-driven networks, UAV applications, and Space-Air-Ground Integrated Networks (SAGIN).

This book deeply integrates FD and ISAC into a complete knowledge system spanning theory, algorithms, and hardware. It covers emerging interdisciplinary research directions combining FD-ISAC with Terahertz, RIS, and AI. Furthermore, it provides extensive hardware implementation data, performance comparison tables, and prototype system design cases, offering significant engineering reference value. In addition, it identifies clear future research directions for currently unresolved problems, helping readers grasp the trajectory of technological evolution.

This book is suitable for the following readership: Researchers and engineers in the fields of communication and sensing, postgraduate students and senior undergraduates in electronic information and communication engineering, industry technical decision-makers engaged in 6G, the Internet of Things (IoT), intelligent transportation, and low-altitude communications.

In summary, this book serves not only as a significant summary of the current state of ISAC research but also as a key reference for propelling the field from theory to practice and from the laboratory to industrialization. We hope this book will provide robust technical support and intellectual inspiration for fostering innovative development in the convergence of 6G communication and intelligent sensing.

Given the author's limited abilities, there are inevitably errors and omissions in this book. Readers are kindly invited to offer criticism and corrections.

Beijing, China Prof. Changhao Du

Acknowledgements First and foremost, we wish to express our deepest gratitude to Mr. Wayne Hu, Senior Editor, Engineering, whose invaluable professional guidance, strategic insights, and unwavering support have been the cornerstone of this book's development. From refining the core framework to navigating the editorial process, his rigorous academic perspective and profound industry expertise have significantly elevated the quality and depth of the work, making his contributions indispensable to the project's success.

We also extend our sincere appreciation to Mr. Rammohan Krishnamurthy, Production Editor, for his meticulous coordination, attentive oversight, and efficient management throughout the entire publication journey. His dedication to ensuring the smooth progression of manuscript polishing, layout design, and logistical coordination has streamlined complex processes and resolved challenges with professionalism, laying a solid foundation for the timely and high-standard release of this book.

Additionally, we would like to acknowledge and thank all the distinguished experts, scholars, and colleagues who have generously devoted their time and efforts to this work. Their constructive feedback, insightful comments on technical content, and selfless sharing of expertise have enriched the book's academic connotations and practical value. This publication is a collective achievement, and we are deeply grateful for the wisdom and support contributed by every individual involved.

Competing Interests The authors have no competing interests to declare that are relevant to the content of this manuscript.

Contents

Part I
Theoretical Foundations of FD-ISAC

Chapter 1
Introduction

Abstract With the rapid advancement of information technology, emerging application scenarios such as smart low-altitude systems, the Internet of Things (IoT), autonomous driving, augmented reality (AR), and virtual reality (VR) are continuously evolving, thereby placing increasing demands on traditional communication systems. Integrated communication and sensing (ISAC) technology facilitates the joint operation of communication and sensing on a single platform, significantly enhancing system performance. The integration of full-duplex (FD) technology further boosts communication capacity and mitigates challenges such as sensing blind spots inherent in half-duplex systems. Consequently, the deep integration of communication and sensing—particularly full-duplex integrated sensing and communication (FD-ISAC) technology—has become a primary focus and breakthrough in current research. This book provides a comprehensive introduction to full-duplex communication-sensing integration, addressing key aspects such as system architecture design, interference elimination, signal processing, communication and sensing algorithms, and hardware implementation. It also offers solutions to the challenges faced by full-duplex ISAC technology. Through systematic discussion, the book aims to provide researchers and engineers with both theoretical foundations and practical technical guidance, thereby promoting the further development and application of full-duplex communication-sensing integration technology.

1.1 The Background and Significance of FD-ISAC

With the rapid development of wireless communication, numerous wireless application scenarios, such as smart factories, low-altitude unmanned systems and vehicle to everything (V2X), are emerging [1]. In smart factory environments, low-latency and high-reliability communication is essential for enabling real-time data exchange among machines, controllers, and cloud platforms. At the same time, high-precision sensing empowers factory nodes and robots to seamlessly navigate, coordinate, and map their surroundings. Moreover, the fusion of communication and sensing not only offers a promising solution for real-time detection, tracking, and classification of non-cooperative unmanned aerial vehicles (UAVs) in low-altitude airspace, but also

C. Du et al., *Full-Duplex Integrated Sensing and Communication Systems*,
https://doi.org/10.1007/978-981-92-0470-0_1

significantly enhances the coordination, control, and navigation capabilities of cooperative UAVs operating within swarms. Furthermore, to enable autonomous driving, vehicles are increasingly equipped with advanced communication transceivers and a wide array of sensors to facilitate both environmental perception and real-time information exchange with roadside units (RSUs), nearby vehicles, and even pedestrians. To this end, these cutting-edge scenarios impose stringent demands on both robust communication performance and precise environmental sensing capabilities.

1.1.1 Disadvantages of Separated Radar and Communication Systems

In the past, communication and sensing modules were typically designed and deployed as independent subsystems, each with its own dedicated hardware, signal processing procedures, and spectrum resources. This separated architecture, while functionally adequate for traditional applications, lacks the efficiency and scalability required by intelligent systems. With the rapidly growing demand for integrated communication-sensing solutions, this fragmented design paradigm has begun to reveal several critical limitations, including increased hardware complexity, higher power consumption, and severe spectrum scarcity [2].

- ***Increased Hardware Complexity:*** The separate design paradigm necessitates redundant hardware components, including dedicated RF front-ends, heterogeneous antenna arrays, and specialized baseband processing units, each tailored for either communication or sensing tasks. The resulting complexity fundamentally conflicts with the miniaturization requirements of B5G/6G network nodes, limiting their deployability in space-constrained environments.
- ***Higher Power Consumption:*** Independent operation of communication and sensing subsystems creates energy inefficiencies due to duplicated signal chains and uncoordinated power management. The concurrent operation of separate RF modules with overlapping functionalities results in redundant power allocation across components such as amplifiers, converters, and oscillators. This significantly limits the energy endurance of the device and poses a major obstacle to the realization of energy-efficient and sustainable green networks.
- ***Severe Spectrum Scarcity:*** The isolated spectrum allocation for communication and sensing functions imposes competing demands on limited frequency resources. As the density of dual-functional devices continues to grow, the use of dedicated, non-overlapping frequency bands for each modality becomes increasingly unsustainable, failing to support concurrent operations and exacerbating spectrum contention. This severely limits the scalability and deployment flexibility of future wireless systems, particularly in spectrum-constrained environments, such as dense industrial networks and vehicular communication scenarios.

Generally speaking, conventional communication-sensing separated architecture is subject to increased hardware complexity, higher power consumption, and severe spectrum scarcity. These impediments significantly hinder the performance optimization and large-scale deployment of communication-sensing systems in emerging application scenarios. Towards that end, there is a strong need to jointly design communication and sensing operations in B5G/6G wireless networks, which motivates the recent research theme of Integrated Sensing and Communications (ISAC).

1.1.2 From Traditional ISAC Systems to FD-ISAC Systems

ISAC technology achieves the convergence and coexistence of communication and sensing by leveraging unified waveforms, multiplexed spectrum, and hardware resources, providing robust support and breakthroughs for emerging information services and applications. Moreover, ISAC can effectively synergize with advanced technologies such as multiple-input multiple-output (MIMO), orthogonal frequency division multiplexing (OFDM), and millimeter-wave (mmWave) technologies, which can substantially enhance the performance of both communication and sensing. Consequently, ISAC has emerged as a central research focus in both academia and industry, demonstrating considerable potential in the development of novel information infrastructures such as integrated air-space networks and intelligent connectivity.

Based on the number of sensing nodes, existing ISAC systems can be systematically classified into three major categories:

- ***Mono-Static Sensing:*** Downlink and uplink signals transmitted by the base station (BS) and user equipment (UE), respectively, are exploited for sensing purposes, while the BS or UE receives the corresponding echo signals reflected from targets using its own receiver. In this configuration, the BS or UE functions as a mono-static radar, for which the transmitter and the receiver are collocated.
- ***Bi-Static Sensing:*** Downlink or uplink signals are reflected off targets and received by another BS, rather than the original transmitter. In this configuration, two spatially separated nodes are jointly involved in the sensing process.
- ***Distributed/Networked Sensing:*** Sensing signals can be emitted by multiple transmitters and received by multiple receivers after being reflected from targets. Both BSs and UEs can function as transmitters or receivers. This corresponds to a multi-static radar system, which requires a certain level of cooperation among the transceivers to enable joint sensing and accurate information fusion.

Among these, mono-static sensing stands out for its simple hardware architecture and inherent synchronization advantages. Unlike bistatic or multistatic setups, it does not require distributed nodes, thereby reducing system complexity. Moreover, since the transmitter and receiver are co-located within the same device, they can share the same local oscillator, effectively eliminating the need for complex time and frequency synchronization mechanisms between separate units. These features make it particularly well-suited for deployment at BSs, referred to as ISAC BSs, where existing infrastructure can be effectively leveraged to support both communication and sensing functionalities in a resource-efficient and cost-effective manner.

However, since the echo signals and the transmitted signals share the same frequency band, ISAC BSs suffer from severe self-interference (SI). The SI power can be up to one hundred billion times stronger than that of the reflected echo signals, leading to a significant degradation in the sensing signal-to-interference-plus-noise ratio (SINR). To address this issue, conventional ISAC BS typically employ time-division duplexing (TDD) schemes, referred to as TDD-ISAC, to avoid severe SI. However, TDD-ISAC systems still exhibit the following limitations:

- ***Communication Capacity Degradation:*** Reserving dedicated time slots for receiving sensing echoes reduces the time available for communication, thereby lowering spectral efficiency (SE) and constraining communication capacity.
- ***Sensing Blind Area:*** The inability to receive sensing echoes during communication time slots significantly hinders effective sensing of the near-field region.
- ***Disruption in Communication and Sensing:*** Since TDD-ISAC systems are required to allocate separate time slots for communication and sensing, they are inherently unable to support the simultaneous operation of both functions. This temporal division constrains the ability of ISAC systems to respond to highly dynamic environmental changes in real-time.

In contrast, FD technology allows ISAC devices to transmit and receive simultaneously, effectively eliminating the sensing blind area and improving the communication capacity [3, 4]. Moreover, FD operation enables seamless communication and sensing without the need for temporal separation, significantly improving ISAC performance in highly dynamic environments. Therefore, if we can combine FD technology with ISAC, refer to as FD-ISAC, the performance of communication and sensing will inevitably be further improved.

1.2 Technical Difficulties and Solutions of FD-ISAC

The FD-ISAC system has problems such as difficulties in signal processing algorithms and hardware implementation, and the overall architecture of FD-ISAC needs to be researched and developed.

To lay the groundwork for FD-ISAC research, we comprehensively detailed the key technologies of ISAC and FD separately. In Chap. 2, focusing on ISAC systems, we examined the core technology of waveform design, elaborating on radar-centric waveforms prioritizing sensing accuracy, communication-centric waveforms optimized for data transmission, and joint waveform designs balancing both objectives. Concurrently, we presented ISAC channel modeling methodologies: communication channels necessitate modeling path loss and small-scale fading, while sensing channels require separate modeling of the target echo channel and environmental clutter channel, establishing the foundation for reliable system operation. In Chap. 3, for FD systems, we analyzed the primary challenge of strong self-interference (SI) arising from simultaneous transmission and reception, compounded by nonlinear distortion and multipath reflections. We investigated solutions to this challenge through a comprehensive tri-domain cancellation approach: spatial domain techniques utilize antenna isolation and beamforming to isolate SI; RF domain methods employ high-linearity amplifiers, optimized front-ends, and analog cancellers to suppress leakage; digital domain strategies leverage adaptive filtering and machine learning algorithms for real-time residual SI cancellation.

Building upon this theoretical foundation, we provided a comprehensive introduction to the key algorithms and implementation methods for ISAC. In Chap. 4, addressing FD-ISAC signal processing, the critical difficulty lies in separating the weak target echo signal from the complex received signal mixture containing residual SI and environmental signals to achieve precise sensing. Echo signal extraction employs Independent Component Analysis (ICA) for blind source separation, multi-tap analog domain cancellation hardware, and Doppler-based dynamic target extraction methods exploiting motion signatures. Subsequently, radar estimation techniques such as periodogram methods, subspace spectrum analysis (e.g., MUSIC), Compressed Sensing (CS), and Deep Learning (DL) are applied to the extracted echoes to obtain sensing information. In Chap. 5, the hardware implementation of FD-ISAC faces challenges related to size, cost, and achieving sufficient SI cancellation performance in integrated systems. Dedicated FD sensing hardware platforms integrating these techniques have been established as testbeds. However, these implementations remain in the early stages, encountering hurdles in miniaturization and cost reduction for widespread adoption. In Chap. 6, the miniaturization and integration of FD-ISAC devices present significant challenges, particularly in designing effective, small-area, low-power on-chip RF cancellers. This chapter details monolithic RFIC solutions classified into four main architectures: Frequency-Domain Equalization (FDE), Time-Domain Equalization (TDE), Hilbert-Transform Equalization (HTE), and MIMO cancellation architectures. Each architecture offers distinct trade-offs in cancellation performance, supported bandwidth, noise figure degradation, and silicon chip area consumption, necessitating future research towards highly reconfigurable, low-power, area-efficient designs.

In conclusion, despite the myriad challenges that FD-ISAC technology faces in architecture design, algorithm development and hardware implementation, these obstacles are progressively being surmounted through advanced research and innovative solutions. Looking ahead, as FD technology continues to mature and undergo optimization, FD-ISAC is poised to play an increasingly critical role in enhancing the synergistic performance of communication and sensing, achieving more efficient resource sharing, bolstering communication security, and reducing system costs. By integrating sophisticated signal processing algorithms with adaptive interference management techniques, FD-ISAC systems will be capable of attaining higher sensing accuracy and improved communication reliability in complex environments. This advancement will not only further propel the development of 5G and 6G networks but also support emerging application domains such as unmanned delivery, intelligent manufacturing, and smart healthcare. Consequently, FD-ISAC will provide robust technical support, fostering the comprehensive and intelligent evolution of the information society.

1.3 Development History of ISAC

1.3.1 Theoretical Developments in Integrated Sensing and Communication Technologies

Over the past seventy years [7], as communication and sensing have transitioned from being separate entities to becoming increasingly integrated, ISAC technology has gradually evolved from its nascent stage to maturity.In particular, recent years have seen a significant push, driven by ISAC-related white papers and relevant protocols from IMT and ITU [8], leading to the recognition of ISAC as a key enabling technology for next-generation wireless networks and radar systems. This recognition has spurred a wave of in-depth research across academia and industry, with the latest advancements already being implemented. The proceeding of ISAC is shown in Fig. 1.1.

The precursor to integrated sensing and communication (ISAC) technology can be traced back to the 1970s, originating from NASA's Space Shuttle program [9], which concurrently employed communication and radar functionalities in space systems. These systems integrated sensing capabilities—including target search and tracking—with bidirectional communication functionalities. Later, in the 1990s, the emergence of the U.S. Advanced Multifunction Radio Frequency Concept (AMRFC) stimulated further theoretical research focused on the integration of communication, sensing, and electronic warfare [10]. Nevertheless, in these early studies, the functionalities for communication and sensing remained relatively distinct.To facilitate radar and communication integration, Kalenichenko et al. [11] introduced the concept of Joint Radar Communication (JRC) in 2006, embedding communication information into linear frequency modulated (LFM) radar signals. Subsequently, further research focused on waveform designs for ISAC, leveraging chirp signals [12, 13], weighted pulse sequences [14], and direct-sequence spread-spectrum (DSSS) techniques [14]. Expanding upon these foundational studies, Chen et al. [15] proposed a novel LFM waveform incorporating minimum shift keying (MSK) modulation, significantly improving its sensing performance.Zhang et al. [16] investigated a compatible waveform design combining linear frequency modulation (LFM) with continuous phase modulation (CPM), effectively mitigating performance degradation caused by power amplifier (PA) nonlinearity and reducing out-of-band interference.However, the high intercept probability associated with linear frequency modulated (LFM)

Fig. 1.1 Proceeding of ISAC

signals poses a significant risk to secure information transmission. Furthermore, ISAC waveforms based on LFM or spread spectrum technologies, characterized by fixed waveform parameters and relatively low data rates, are insufficient for large-scale data transmission requirements.Orthogonal frequency division multiplexing (OFDM) waveforms were initially introduced for radar sensing applications in 2009 [17], demonstrating the feasibility of simultaneously achieving communication and sensing functionalities at 24 GHz. In this approach, high-speed data streams were partitioned into multiple subcarriers for parallel transmission, thereby validating the integration capability of OFDM waveforms. Additionally, the inherent resistance of OFDM to multipath fading has been leveraged in ISAC scenarios involving multipath and multiple users [18, 19]. Subsequent studies have extensively verified and optimized both the communication and sensing performance of OFDM-based systems [20–22].

Multiple-input multiple-output (MIMO) is another critical technology for enabling integrated sensing and communication (ISAC), facilitating communication with multiple users and sensing multiple targets using multiple beams [23]. In 2010, the concept of phased-array MIMO radar was introduced [24], where the transmit array is divided into multiple overlapping subarrays that transmit orthogonal waveforms, thereby enhancing beamforming performance.This development inspired further research into MIMO-ISAC systems in subsequent years. Li et al. [25] proposed a coexistence scheme for MIMO radar and communication systems, which maximizes the signal-to-interference-plus-noise ratio (SINR) at the receiver while simultaneously achieving specific sensing functions.BouDaher et al. [26] introduced a novel information embedding technique by embedding symbols into pulse waveforms, providing the theoretical foundation for subsequent research. Additionally, the MIMO radar systems based on matrix completion (MIMO-MC) [27–29] minimized mutual interference while sharing the spectrum, thereby alleviating the constraints of radar systems on communication power and speed.Qian et al. [30] investigated beamforming schemes for MIMO communication radar systems and performed joint parameter optimization based on radar system filter design. To reduce complexity while maintaining the same performance, Arik et al. [31] separated the antennas for communication and radar functions and proposed a weighted optimization algorithm for their shared deployment.

Moreover, millimeter-wave (mmWave) [32] technology is well suited for integration with massive MIMO (mMIMO) due to its high beamforming gain [33], which has spurred research into ISAC technologies based on mmWave-mMIMO systems.In 2015, Guerra et al. [34] investigated the limits of positioning and orientation performance in millimeter-wave wideband large-scale array networks, deriving the effects of array structure, bandwidth, and synchronization errors on positioning accuracy.Shahmansoori et al. [35] studied the Cramér-Rao bound (CRB) for positioning and rotation angle estimation uncertainty in mmWave-MIMO systems. Furthermore, the scenario was extended to non-line-of-sight (NLoS) conditions and scattering, and the positioning capability of mmWave signals under different multipath delays was analyzed [36].In addition to enhancing sensing performance, the reciprocity of channel estimation in mMIMO-ISAC technology [37] enables the integration of radar antenna array motion parameters and echoes to obtain channel state information. As a result, combining the narrow beams and large array codebook characteristics of mmWave-mMIMO systems with joint ISAC technology effectively reduces beam training overhead and addresses channel estimation challenges in dynamic scenarios [38].

In 2016, an ISAC system based on MIMO-OFDM was proposed [39]. In this MIMO-OFDM system, each antenna transmits an integrated OFDM waveform across multiple sub-bands, making full use of the array structure and signal bandwidth, while also providing high-resolution angle and distance estimation capabilities.To address the high peak-to-average power ratio (PAPR) issue of OFDM signals, Bai et al. [40] optimized the MIMO-OFDM waveform and system design. By incorporating complex orthogonal design, they achieve code-domain orthogonality, effectively suppressing sidelobes and improving distance resolution.Liu et al. [41] formulated an optimization problem, highlighted the advantages of MIMO-OFDM over single-input single-output (SISO) systems, and conducted a performance analysis. In 2021, Keskin et al. [42] further examined high-mobility scenarios and analyzed carrier interference (ICI) issues in OFDM systems. The sensing algorithm of the proposed MIMO-OFDM JRC system demonstrated superior detection and estimation performance. Meanwhile, Johnston et al. [43] addressed multi-user requirements by designing waveforms and receiving filters.

Since 2020, theoretical research on ISAC integrated with other emerging technologies has gained increasing attention. For example, the combination of ISAC and reconfigurable intelligent surface (RIS) technology [44] enhances energy efficiency (EE) while improving sensing accuracy and coverage.Shao et al. [45] utilized reconfigurable intelligent surface (RIS) technology for signal detection and combined it with multi-signal classification algorithms to sense the direction of nearby targets, thereby maximizing the average total power of the echo signals at the IRS sensor. Sankar et al. [46] divided the RIS reflection units into two groups for communication and sensing beams, leading to improved spectral efficiency (SE).In 2024, Hua et al. [47] utilized reconfigurable intelligent surface (RIS) technology to achieve target sensing and localization in non-line-of-sight (NLoS) links, with the proposed architecture achieving sub-meter localization accuracy over extended distances. Research on orthogonal time frequency space (OTFS) technology [48–50], based on improved OFDM waveforms, has demonstrated superior sensing performance in high-speed scenarios. Additionally, integrated sensing and communication (ISAC) has significant applications and can be combined with low Earth orbit satellite networks [51], terahertz technology [52], and digital twins [53, 54].

Since 2018, numerous studies on FD-ISAC have emerged. Hassani et al. [55] developed an architecture based on an electrical balance duplexer (EBD) that can balance radar communication capabilities while mitigating SI. In 2019, Barneto et al. [56] developed an FD-ISAC architecture based on OFDM. This architecture enabled the sensing of both static and moving targets, such as drones and vehicles. Hassani et al. [57] conducted a FD-ISAC scenario simulation based on IEEE 802.11ac signals, which demonstrated sensing capabilities sufficient for gesture recognition and tracking applications. Furthermore, Islam et al. [58] applied FD-ISAC to mmWave systems, performing joint optimization for beamforming and SIC, which enabled the estimation of sensing information for multiple radar targets. Zhang et al. [59] designed a FD Integrated Access and Backhaul (IAB) network, which effectively improved node spectrum efficiency (SE) and sensing accuracy. Finally, Duan et al. [60] proposed a novel frequency-domain differential SIC scheme, which is effectively applicable to FD-ISAC systems and can eliminate dynamic interference while extracting echo signals from moving targets.

1.3.2 Experimental Verification of ISAC

Radar sensing and communication hardware systems have developed independently for a long period. With the advent of ISAC, radar and communication now coexist. The hardware prototype design for ISAC systems needs to integrate theoretical frameworks. Considering factors such as hardware compatibility, cost, size, and power consumption, rational design of radio frequency (RF) and baseband transceiver structures is required. Depending on the duplexing method of ISAC systems, hardware validation is classified into TDD architecture and FD architecture.

The experimental verification of TDD-ISAC systems began in 2006 when Northrop Grumman tested the transmission data performance of their Active Phased Array Radar (AESA) [61], initially demonstrating the communication performance of their prototype. In 2011, [70] validated the performance of ISAC systems based on OFDM and DSSS waveforms and tested their operability in real road scenarios using a dedicated system demonstrator, though it could not resolve moving objects. Further, in 2012, Sturm et al. [62] tested signals parameterized according to IEEE 802.11a/p standards using USRP as the frontend, with their ISAC system capable of resolving moving objects. In 2017, the University of Kansas [63] completed prototype validation of ISAC systems using LFM waveforms in an indoor anechoic chamber. They confirmed that their platform demonstrated good communication performance and matched the expected radar waveform through beam pattern measurements, radar waveform validation, and communication beam error rate analysis. This was the first test and validation of ISAC waveforms. In 2019, Wang et al. [64, 65] applied ISAC technology to SAR radar, demonstrating its imaging capabilities, and subsequently conducted experiments on high-speed data communication [66]. In the same year, Ravenscroft et al. [67] evaluated frequency modulated radar waveforms with OFDM subcarriers and achieved a receiver signal-to-noise ratio (SNR) of approximately 20 dB. To achieve higher communication rates, in 2021, Zhang et al. [68] combined 5G mmWave technology with ISAC for vehicular networks, achieving radar ranging accuracy of 0.25 m and a data rate of 2.8 Gbps. Additionally, Ma et al. [69] implemented an ISAC system based on Generalized Spatial Modulation (GSM), achieving higher communication rates by embedding additional data bits in antenna selection and improving angular resolution while reducing sidelobe levels in the beam pattern through spatial diversity induced by GSM transmission. In 2022, Xu et al. [70] designed an SDR testbed based on MIMO-OFDM, which demonstrated the capability to count individuals entering a room with high accuracy. To further enhance sensing accuracy, Sakhnini et al. [71] designed and validated the performance of an mMIMO-OFDM testbed, achieving centimeter-level high-precision localization of moving objects and high communication capacity. As research frequencies advanced, in 2024, Ozkaptan et al. [72] prototyped a 24GHz ISAC system based on an mmWave-MIMO testbed and evaluated the performance of radar-assisted precoding methods, achieving both data transmission and high-resolution radar processing capabilities. In the same year, Ouyang et al. [73] developed and prototyped a testbed for OTFS-ISAC signals.

The experimental verification of FD-ISAC systems began in 2018, with Hassani et al. [55] from KU Leuven proposing an FD transceiver architecture with Doppler radar detection capabilities. In 2019, Barneto et al. [56] addressed the challenges and solutions of ISAC base station (BS) with both 4G LTE and 5G NR OFDM waveforms. Moreover, in 2021, The University of Leuven built a prototype based on a FD OFDM architecture with a low frequency band of 1–2.4 GHz and a narrow bandwidth of 5-40 MHz in the delay concentration, and a SIC capability of 86 dB, which achieved a detection range of up to 20 m and a moving velocity detection of 62.2 m/s [74, 75], with a velocity resolution of 8.6 cm/s. Furthermore, the University of Toulouse's Mercier et al. designed a weighted CP-OFDM for radar sensing algorithms [76]. In addition, Zhang et al. [77] from the University of Edinburgh used an OTFS hybrid waveform to achieve a detection range of 20 m and a detection speed of 38 m/s at a carrier frequency of 28 GHz, a bandwidth of 50 MHz, and a transmit power of 20dBm. Although FD-ISAC hardware platforms is advancing toward higher frequency bands, with improving sensing performance, most systems can only detect basic information such as target distance and speed, lack imaging capabilities, and research on wideband and on-chip solutions remains limited.

1.3.3 Application Scenarios of ISAC

ISAC technology enables the simultaneous execution of functions such as distance measurement, positioning, imaging, and environment reconstruction while communicating. By leveraging the obtained sensing information, the communication performance of the system can be improved to address varying service demands across different scenarios. Consequently, ISAC systems are applicable in domains such as the Internet of Everything (IoE) and smart cities, delivering enhanced and efficient information services. The following will elaborate on several representative ISAC application scenarios,as shown in Fig. 1.2.

In smart transportation scenarios, the use of ISAC BSs or collaboration through vehicle-to-infrastructure systems [78] enables comprehensive, all-weather, and continuous detection of vehicle positions, driving trajectories, and speed information, with the sensing measurement data uploaded to a processing center. This capability significantly enhances the intelligent monitoring of highway operational states and provides robust data support for road supervision. Moreover, ISAC BSs can achieve

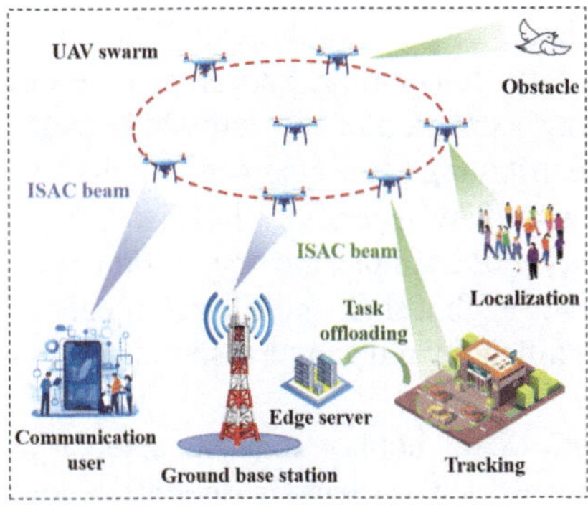

Fig. 1.2 Application Scenarios of ISAC

road environment perception. This approach effectively constructs high-precision maps, providing beyond-visual-range assistance for autonomous vehicles [79–81]. In June 2024, ZTE, in collaboration with Shanghai Telecom, demonstrated the first 5G-A BS supporting both 10G experience and ISAC services. This demonstration achieved a combination of 4K High Definition live broadcasting and real-time channel monitoring, supporting capacity assurance in the Lujiazui hotspot area and smart governance of the Jiangpu region [83].

In smart living scenarios, intelligent IoT devices should possess the capability to sense nearby users' physical and behavioral characteristics, enabling the detection of motion information in surrounding indoor environments [82, 84]. Furthermore, based on the location and status sensing of home devices, remote control and intelligent management of smart homes can be achieved [85].

Simultaneously, by sensing physiological parameters of the human body, such as heart rate and respiration, real-time monitoring and alerting of user health conditions can be performed [86]. In November 2022, ZTE conducted a respiration sensing test in an indoor environment using a 4.9 GHz low-frequency 5G commercial BS [87]. The respiration monitoring capability in line-of-sight (LoS) scenarios is comparable to commercial respiration monitoring instruments, which can support applications in health management and other sensing areas. In 2024, NTT DoCoMo in Japan [88] initiated a trial project deploying ISAC BSs and sensor devices within a park to monitor environmental parameters (e.g., temperature, humidity, air quality), personnel activity, and equipment operating status. The sensed data is transmitted in real-time to a management center via communication networks for environmental management, resource optimization, and security monitoring.

In intelligent low-altitude scenarios, ISAC technology can provide reliable communication services to ground and low-altitude users while performing environmental sensing that can assists in trajectory planning and conserve resources [89]. Moreover, unmanned aerial vehicles (UAVs) with ISAC technology can detect and sense key information such as the position, speed, and trajectory of drones [90], monitor unauthorized flights, prevent air collisions, and manage low-altitude drone operations for enhanced security. In June 2024, China Telecom [91] developed the first 5G+ drone smart airport platform, completing validation of low-altitude logistics, security, and route protection in Shenzhen, contributing to the emergence of the low-altitude economy. The integration of drones with ISAC technology has also brought transformative changes to the logistics industry. For example, the first 100 km 5G-A ISAC cross-sea route network coverage between Zhoushan and Shanghai enables drones to transport goods such as seafood, significantly reducing transportation time compared to traditional cargo ships [92].

As an innovative technological concept, ISAC holds extensive application prospects. It has widespread applications in areas such as smart transportation, low-altitude intelligence, smart homes, and industrial IoT, offering more efficient and intelligent solutions for daily life and work. FD-ISAC has distinct advantages over TDD mode, being more suitable for high-throughput and precise sensing scenarios. Although theoretical research has started later and there are still many technical challenges to overcome, and experimental validation remains immature, this situation further motivates continued in-depth research by universities and related institutions, and promote the continuous growth of commercial applications.

1.4 How This Book Is Organized

The Section structure of this paper is shown in Fig. 1.3. Section 2 introduces the ISAC system model, channel model and beamforming design of ISAC. After that,

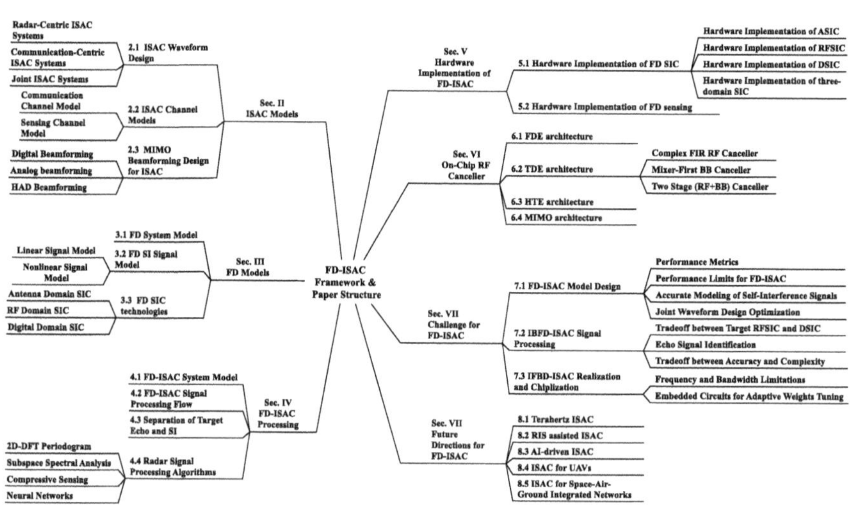

Fig. 1.3 Structure of the Paper

Sect. 3 presents the system and SIC model of FD-ISAC. Moreover, Sect. 4 presents the FD-ISAC processing technology, including Processing Flow and Radar Signal Processing Algorithms. Furthermore, Sect. 5 and 6 introduces the hardware and on-chip implementation of FD-ISAC.Finally, Sects. 7 and 8 analyses the challenges and future directions in the field.

References

1. Masouros C, Zhang JA, Liu F et al (2023) Guest editorial: integrated sensing and communications for 6G. IEEE Trans Wirel Commun 30(1):14–15
2. Gong Y, Li X, Meng F et al (2024) Toward green RF chain design for integrated sensing and communications: technologies and future directions. IEEE Trans Commun 62(9):36–42
3. Xiao Z, Zeng Y (2022) Waveform design and performance analysis for full-duplex integrated sensing and communication. IEEE J Sel Areas Commun 40(6):1823–1837
4. Tang A, Wang X, Zhang JA (2024) Interference management for full-duplex ISAC in B5G/6G networks: architectures, challenges, and solutions. IEEE Trans Commun 62(9):20–26
5. Ren J, Zhang N, Gao Y et al (2020) Guest editorial: service-oriented space-air-ground integrated networks. IEEE Trans Wirel Commun 27(6):10–11
6. Liu J, Shi Y, Fadlullah ZM et al (2018) Space-air-ground integrated network: a survey. IEEE Commun Surv Tutor 20(4):2714–2741
7. Liu F, Zheng L, Cui Y et al (2023) Seventy years of radar and communications: the road from separation to integration. IEEE Signal Process Mag 40(5):106–121
8. Kaushik A, Singh R, Dayarathna S et al (2024) Toward integrated sensing and communications for 6G: key enabling technologies, standardization, and challenges. IEEE Commun Stand Mag 8(2):52–59
9. Cager R, LaFlame D, Parode L (1978) Orbiter ku-band integrated radar and communications subsystem. IEEE Trans Commun 26(11):1604–1619
10. Hughes P K, Choe J Y (2000) Overview of advanced multifunction RF system (AMRFS). In: Proc IEEE international symposium phased array system and technology, pp 21–24
11. Kalenichenko S P, Mikhailov V N (2006) The joint radar targets detecting and communication system. In: Proceedings of the international radar symposium, pp 1–4
12. Roberton M, Brown ER (2003) Integrated radar and communications based on chirped spread-spectrum techniques. Proc IEEE MTT-S Int Microwave Symp Dig 1:611–614
13. Zhang C, Chen Q (2013) Design of signal-sharing for radar and communication. In: Proceedings of international conference on mechatronics science electric engineering computing, pp 1250–1253
14. Jamil M, Zepernick H J, Pettersson M I (2008) On integrated radar and communication systems using Oppermann sequences. In: Proceedings of military communications conference, pp 1–6
15. Chen X, Wang X, Xu S et al (2011) A novel radar waveform compatible with communication. In: Proceedings of international conference on computing problem-solving, pp 177–181
16. Zhang Y, Li Q, Huang L et al (2017) Waveform design for joint radar-communication with nonideal power amplifier and outband interference. In: Proceedings of IEEE wireless communication network conference, pp 1–6
17. Sturm C, Zwick T, Wiesbeck W (2009) An OFDM system concept for joint radar and communications operations. In: Proceedings of IEEE vehicular technology conference, pp 1–5
18. Sit Y L, Reichardt L, Sturm C et al (2011) Extension of the OFDM joint radar-communication system for a multipath, multiuser scenario. In: Proceedings of IEEE nature radar conference, pp 718–723
19. Sit Y L, Sturm C, Zwick T (2012) One-stage selective interference cancellation for the OFDM joint radar-communication system. In: Proceedings of German microwave conference, pp 1–4

20. Gogineni S, Rangaswamy M, Nehorai A (2013) Multi-modal OFDM waveform design. In: Proceedings of IEEE nature radar conference, pp 1–5
21. Bica M, Huang K W, Koivunen V et al (2016) Mutual information based radar waveform design for joint radar and cellular communication systems. In: Proceedings of IEEE international conference on acoustic speech signal process, pp 3671–3675
22. Liu Y, Liao G, Xu J et al (2017) Adaptive OFDM integrated radar and communications waveform design based on information theory. IEEE Commun Lett 21(10):2174–2177
23. Liu F, Masouros C, Li A et al (2018) MU-MIMO communications with MIMO radar: from co-existence to joint transmission. IEEE Trans Wirel Commun 17(4):2755–2770
24. Hassanien A, Vorobyov SA (2010) Phased-MIMO radar: a tradeoff between phased-array and MIMO radars. IEEE Trans Signal Process 58(6):3137–3151
25. Li B, Kumar H, Petropulu A P (2016) A joint design approach for spectrum sharing between radar and communication systems. In: Proceedings of IEEE international conference on acoustic speech signal process, pp 3306–3310
26. BouDaher E, Hassanien A, Aboutanios E et al (2016) Towards a dual-function MIMO radar-communication system. In: Proc IEEE radar conference, pp 1–6
27. Li B, Petropulu A (2015) Spectrum sharing between matrix completion based MIMO radars and a MIMO communication system. In: Proceedings of IEEE international conference on acoustic speech signal process, pp 2444–2448
28. Li B, Petropulu AP (2017) Joint transmit designs for coexistence of MIMO wireless communications and sparse sensing radars in clutter. IEEE Trans Aerosp Electron Syst 53(6):2846–2864
29. Li B, Petropulu AP, Trappe W (2016) Optimum co-design for spectrum sharing between matrix completion based MIMO radars and a MIMO communication system. IEEE Trans Signal Process 64(17):4562–4575
30. Qian J, Lops M, Zheng L et al (2018) Joint system design for coexistence of MIMO radar and MIMO communication. IEEE Trans Signal Process 66(13):3504–3519
31. Arik M, Akan OB (2019) Realizing joint radar-communications in coherent MIMO radars. Phys Commun 32:145–159
32. Marzetta TL (2010) Noncooperative cellular wireless with unlimited numbers of base station antennas. IEEE Trans Wirel Commun 9(11):3590–3600
33. Rappaport TS, Sun S, Mayzus R et al (2013) Millimeter wave mobile communications for 5G cellular: it will work! IEEE Access 1:335–349
34. Guerra A, Guidi F, Dardari D (2015) Position and orientation error bound for wideband massive antenna arrays. In: Proceedings of IEEE international conference on communication workshop, pp 853–858
35. Shahmansoori A, Garcia G E, Destino G et al (2015) 5G position and orientation estimation through millimeter wave MIMO. In: Proceedings of IEEE Globecom workshops, pp 1–6
36. Shahmansoori A, Garcia GE, Destino G et al (2018) Position and orientation estimation through millimeter-wave MIMO in 5G systems. IEEE Trans Wirel Commun 17(3):1822–1835
37. Zhang JA, Huang X, Guo YJ et al (2019) Multibeam for joint communication and radar sensing using steerable analog antenna arrays. IEEE Trans Veh Technol 68(1):671–685
38. Zhang J, Huang Y, Zhou Y et al (2020) Beam alignment and tracking for millimeter wave communications via bandit learning. IEEE Trans Commun 68(9):5519–5533
39. Liu Y, Liao G, Yang Z (2016) Range and angle estimation for MIMO-OFDM integrated radar and communication systems. In: Proceedings of CIE international conference on radar, pp 1–4
40. Bai T, Hu H, Song X (2017) OFDM MIMO radar waveform design with high range resolution and low sidelobe level. In: Proceedings of international conference on communication and technology, pp 1065–1069
41. Liu Y, Liao G, Yang Z et al (2017) Design of integrated radar and communication system based on MIMO-OFDM waveform. J Syst Eng Electron 28(4):669–680
42. Keskin MF, Wymeersch H, Koivunen V (2021) MIMO-OFDM joint radar-communications: is ICI friend or foe? IEEE J Sel Top Signal Process 15(6):1393–1408
43. Johnston J, Venturino L, Grossi E et al (2022) MIMO OFDM dual-function radar-communication under error rate and beampattern constraints. IEEE J Sel Areas Commun 40(6):1951–1964

44. Di Renzo M, Zappone A, Debbah M et al (2020) Smart radio environments empowered by reconfigurable intelligent surfaces: how it works, state of research, and the road ahead. IEEE J Sel Areas Commun 38(11):2450–2525
45. Shao X, You C, Ma W et al (2022) Target sensing with intelligent reflecting surface: architecture and performance. IEEE J Sel Areas Commun 40(7):2070–2084
46. Prasobh Sankar R S, Deepak B, Chepuri S P (2021) Joint Communication and Radar Sensing with Reconfigurable Intelligent Surfaces. In: Proceedings of IEEE workshop signal process advanced wireless communication, pp 471–475
47. Hua M, Wu Q, Chen W et al (2024) Intelligent reflecting surface-assisted localization: performance analysis and algorithm design. IEEE Wirel Commun Lett 13(1):84–88
48. Gaudio L, Kobayashi M, Caire G et al (2020) On the effectiveness of OTFS for joint radar parameter estimation and communication. IEEE Trans Wirel Commun 19(9):5951–5965
49. Xia X, Xu K, Wang Y et al (2024) Achieving better accuracy with less computations: a delay-doppler spectrum matching assisted active sensing framework for OTFS based ISAC systems. IEEE Trans Wirel Commun 23(6):6204–6220
50. Li S, Yuan W, Liu C et al (2022) A novel ISAC transmission framework based on spatially-spread orthogonal time frequency space modulation. IEEE J Sel Areas Commun 40(6):1854–1872
51. Kodheli O, Lagunas E, Maturo N et al (2021) Satellite communications in the new space era: a survey and future challenges. IEEE Commun Surv Tutor 23(1):70–109
52. Han C, Wu Y, Chen Z et al (2024) THz ISAC: a physical-layer perspective of terahertz integrated sensing and communication. IEEE Trans Commun 62(2):102–108
53. Wei Z, Du Y, Zhang Q et al (2024) Integrated sensing and communication driven digital twin for intelligent machine network. IEEE Internet Things Mag 7(4):60–67
54. Gong Y, Wei Y, Feng Z et al (2023) Resource Allocation for Integrated Sensing and Communication in Digital Twin Enabled Internet of Vehicles. IEEE Trans Veh Technol 72(4):4510–4524
55. Hassani S A, Guevara A, Parashar K et al (2018) An in-band full-duplex transceiver for simultaneous communication and environmental sensing. In: Proceedings of asilomar conference signals system computing, pp 1389–1394
56. Baquero Barneto C, Riihonen T, Turunen M et al (2019) Full-duplex OFDM radar with LTE and 5G NR waveforms: challenges, solutions, and measurements. IEEE Trans Microw Theory Tech 67(10):4042–4054
57. Hassani S A, van Liempd B, Bourdoux A et al (2021) Adaptive filter design for simultaneous in-band full-duplex communication and radar. In: Proceedings of European radar conference, pp 5–8
58. Islam M A, Alexandropoulos G C, Smida B (2022) Integrated Sensing and Communication with Millimeter Wave Full Duplex Hybrid Beamforming. In: Proceedings of IEEE international conference on communication, pp 4673–4678
59. Zhang J, Garg N, Ratnarajah T (2022) Design of In-Band-Full-Duplex IAB Networks for Integrated Sensing and Communications. In: Proceedings of IEEE international conference on communication, pp 1888–1893
60. Duan B, Chen C, Pan W et al (2024) Frequency-domain differential interference cancellation for full-duplex OFDM ISAC systems. IEEE Trans Veh Technol 73(6):8615–8631
61. Du X, Ke M (2009) Design of T/R-module for the ultra-wide-band multi-function AESA. In: Proceedings of Asia-Pacific conference synth aperture radar, pp 261–262
62. Sturm C, Wiesbeck W (2011) Waveform design and signal processing aspects for fusion of wireless communications and radar sensing. Proc IEEE 99(7):1236–1259
63. McCormick P M, Ravenscroft B, Blunt S D et al (2017) Simultaneous radar and communication emissions from a common aperture, Part II: experimentation. In: Proceedings of IEEE radar conference, pp 1697–1702
64. Wang J, Liang X, Chen L et al (2019) Joint wireless communication and high resolution SAR imaging using airborne MIMO radar system. In: Proceedings of IEEE international geoscience remote sensor symposium, pp 2511–2514

65. Wang J, Liang XD, Chen LY et al (2019) First demonstration of joint wireless communication and high-resolution SAR imaging using airborne MIMO radar system. IEEE Trans Geosci Remote Sens 57(9):6619–6632
66. Wang J, Liang XD, Chen LY et al (2022) First demonstration of airborne MIMO SAR system for multimodal operation. IEEE Trans Geosci Remote Sens 60:1–13
67. Ravenscroft B, McCormick P M, Blunt S et al (2019) Experimental assessment of tandem-hopped radar and communications (THoRaCs). In: Proceedings of international radar conference, pp 1–6
68. Zhang Q, Wang X, Li Z et al (2021) Design and performance evaluation of joint sensing and communication integrated system for 5G mm wave enabled CAVs. IEEE J Sel Top Signal Process 15(6):1500–1514
69. Ma D, Shlezinger N, Huang T et al (2021) Spatial modulation for joint radar-communications systems: design, analysis, and hardware prototype. IEEE Trans Veh Technol 70(3):2283–2298
70. Xu T, Liu F, Masouros C et al (2022) An experimental proof of concept for integrated sensing and communications waveform design. IEEE Open J Commun Soc 3:1643–1655
71. Sakhnini A, De Bast S, Guenach M et al (2022) Near-field coherent radar sensing using a massive MIMO communication testbed. IEEE Trans Wirel Commun 21(8):6256–6270
72. Ozkaptan CD, Zhu H, Ekici E et al (2024) A mm wave MIMO joint radar-communication testbed with radar-assisted precoding. IEEE Trans Wirel Commun 23(7):7079–7094
73. Ouyang M, Wang Y, Zhang T et al (2024) An RFSoC-Based Scalable ISAC Prototyping Platform with OTFS Waveform. In: Proceedings of IEEE wireless communication network conference, pp 1–5
74. Hassani SA, Lampu V, Parashar K et al (2021) In-band full-duplex radar-communication system. IEEE Syst J 15(1):1086–1097
75. Hassani SA, van Liempd B, Bourdoux A et al (2022) Joint in-band full-duplex communication and radar processing. IEEE Syst J 16(2):3391–3399
76. Mercier S, Roque D, Bidon S (2019) Study of the target self-interference in a low-complexity OFDM-based radar receiver. IEEE Trans Aerosp Electron Syst 55(3):1200–1212
77. Zhang J, Ratnarajah T (2023) Air-ground OTFS in-band-full-duplex integrated sensing and communications systems. In: Proc IEEE international conference on communication, pp 2637–2642
78. Zeng T, Semiari O, Saad W et al (2019) Joint communication and control for wireless autonomous vehicular platoon systems. IEEE Trans Commun 67(11):7907–7922
79. Baquero Barneto C, Rastorgueva-Foi E, Keskin MF et al (2022) Millimeter-wave mobile sensing and environment mapping: models, algorithms and validation. IEEE Trans Veh Technol 71(4):3900–3916
80. Ge Y, Kaltiokallio O, Kim H et al (2022) A computationally efficient EK-PMBM filter for bistatic mmwave radio SLAM. IEEE J Sel Areas Commun 40(7):2179–2192
81. Yang J, Wen CK, Jin S (2022) Hybrid active and passive sensing for SLAM in wireless communication systems. IEEE J Sel Areas Commun 40(7):2146–2163
82. Gao K, Wang H, Lv H et al (2022) Toward 5G NR high-precision indoor positioning via channel frequency response: a new paradigm and dataset generation method. IEEE J Sel Areas Commun 40(7):2233–2247
83. Zhongxing Telecom Equipment (2024) Shanghai telecom partners with ZTE to Create 5G-A pass-through all-in-one ar live streaming debut. Available via ZTE. https://www.zte.com.cn/china/about/news/20240627c2.html
84. Zeng Y, Wu D, Xiong J et al (2020) MultiSense: enabling multi-person respiration sensing with commodity wifi. Proc ACM Interact Mob Wearable Ubiquitous Technol 4(3):102
85. Jiang H, Cai C, Ma X et al (2018) Smart home based on wifi sensing: a survey. IEEE Access 6:13317–13325
86. Li X, Cui Y, Zhang JA et al (2023) Integrated human activity sensing and communications. IEEE Trans Commun 61(5):90–96
87. Zhongxing Telecom Equipment (2022) ZTE completes industry's first verification of pass-through convergence tests based on low-frequency 5G commercial base stations. Available via ZTE. https://www.zte.com.cn/china/about/news/20221117c1.html

88. DoCoMo Japan (2024) The wireless technology division is currently working on 6G pre-research for the 2030s. Available via DoCoMo Labs. http://www.docomolabs.com.cn/research/index.html
89. Cui Y, Feng Z, Zhang Q et al (2023) Toward trusted and swift UAV communication: ISAC-enabled dual identity mapping. IEEE Trans Wirel Commun 30(1):58–66
90. Wang J, Varshney N, Gentile C et al (2022) Integrated sensing and communication: enabling techniques, applications, tools and data sets, standardization, and future directions. IEEE Internet Things J 9(23):23416–23440
91. China Telecom Corp Ltd (2024) Low-altitude economic applications spread their wings with the help of 5G. Available via the paper. https://www.thepaper.cn/newsDetail_forward_27694813
92. China Mobile Communications Group Co Ltd (2024) Zhoushan-Shanghai realizes low altitude coverage of 100 kilometers of 5G-a pass-through sensory integration cross-sea route. Available via SASAC. http://www.sasac.gov.cn/n2588025/n2588124/c30824040/content.html

Chapter 2
ISAC Models

Abstract Integrated Sensing and Communication (ISAC) is defined as a design approach and enabling technology that combines sensing and communication functions to achieve efficient utilization of wireless resources, thereby mutually benefiting both functions. The communication and sensing functions can be co-implemented on the same waveform through well-designed waveforms, which further facilitates efficient spectrum reuse and improves hardware integration. Additionally, accurate channel modeling provides fundamental knowledge for both reliable communication and precise sensing in ISAC system. This chapter therefore provides an overview of ISAC waveform design and channel modeling.

2.1 ISAC Waveform Design

The design of the ISAC signal waveform, as the foundation of ISAC technology, impacts the sensing and communication performance directly [1]. In order to satisfy different performance requirements, ISAC systems can be categorized into three main types, each distinguished by the underlying design of the waveforms, as shown in Fig. 2.1. The first type is the radar-centric ISAC system, which focuses primarily on utilizing modulation of the radar's waveform to carry communication information. Such systems tend to focus on environmental sensing at the expense of communication efficiency. The second category is communication-centric ISAC systems, which generally use a multicarrier design to realize both communication and sensing functions. In this case, the waveform design prioritizes the communication function, followed by the sensing performance. Finally, the goal of joint ISAC systems is to achieve a balance between sensing and communication functions by integrating them into a unified waveform. These systems require advanced signal processing techniques and efficient waveform design to achieve effective coexistence of the two functions without significant performance degradation.

This section provides an insight into the models that characterize these three ISAC system types, highlighting their unique features and fundamentals.

C. Du et al., *Full-Duplex Integrated Sensing and Communication Systems*,
https://doi.org/10.1007/978-981-92-0470-0_2

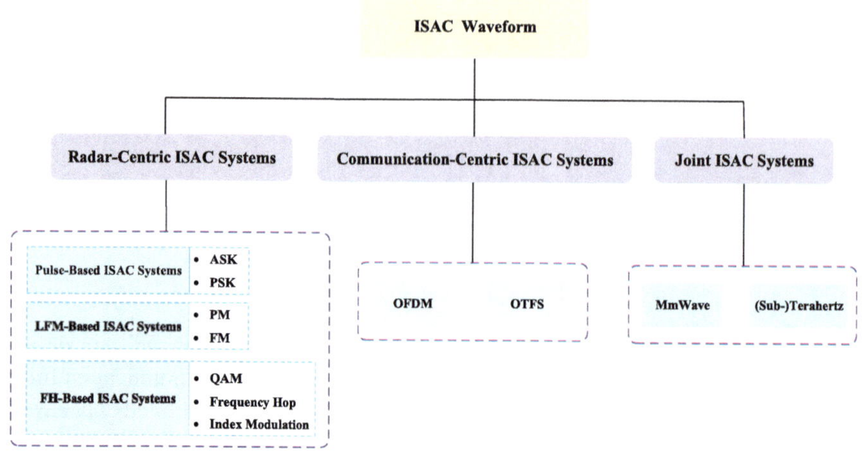

Fig. 2.1 Branch diagram of ISAC Waveform Designs

2.1.1 Radar-Centric ISAC Systems

In radar-centric ISAC systems, communication is enabled within an existing radar framework by embedding communication signals into radar waveforms. Early radar-centric ISAC systems, as demonstrated in works by [2], have shown feasibility in both static and mobile radar setups, where radar communication networks can be built. A key advantage of radar-centric systems is the ability to achieve long-range communication due to radar's inherent wide-range sensing capabilities. However, the data rate is often limited due to the constraints of radar waveforms. There are mainly three categories for radar-centric ISAC systems, i.e., beampattern modulation and index modulation.

(1) Beampattern Modulation ISAC Systems: In beampattern modulation ISAC systems, communication information is embedded within radar pulses using beampattern modulation techniques. This type of information embedding uses exactly the same waveform for radar and communications. However, depending on the transmitted information, it is possible to change the radar beam pattern in the amplitude and phase of the side lobes [3–5].

Consider a SIMO radar system consisting of M antennas. Denote the total transmit power by P. The baseband transmitted signal during the τth radar pulse can be defined as

$$\mathbf{s}(t;\tau) = \sqrt{P}\mathbf{w}^{*}\nu(t), \tag{2.1}$$

where t and τ represent fast time and pulse index, respectively, the notation $(\cdot)^{*}$ denotes the complex conjugate operation, $\mathbf{w}$ refers to the beamforming weight vector with unit norm, and $\nu(t)$ represents the radar waveform. Assuming the radar signal

$\nu(t)$ has unit energy, we have $\int_{T_\nu} |\nu(t)|^2 dt = 1$, where T_ν denotes the duration of waveform.

By modulating the radar beam pattern, communication information can be embedded into the radar signal. The modulated beam pattern during the radar pulse becomes

$$\begin{aligned} G(\theta) &= \mathbf{a}^T(\theta)\mathbb{E}\{\mathbf{s}(t;\tau)\mathbf{s}^H(t;\tau)\}\mathbf{a}(\theta) \\ &= P|\mathbf{w}^H\mathbf{a}(\theta)|^2 = P|g(\theta)|^2, \end{aligned} \tag{2.2}$$

where θ denotes the spatial angle, $\mathbf{a}(\theta)$ denotes the transmit array guidance vector, $\mathbb{E}\{\cdot\}$ denotes the expectation operator, and $g(\theta) = \mathbf{w}^H\mathbf{a}(\theta)$ denotes the complex response of the beamformer in the direction θ.

By designing different $\mathbf{w}_n$ to achieve sidelobe amplitude encoding, for example:

$$P\left|\mathbf{w}_n^H\mathbf{a}(\theta_c)\right|^2 = \delta_n, \quad n = 1, 2, \ldots, N_{\text{AM}}$$

where θ_c is the communication direction, $\{\delta_n\}$ denotes the amplitude symbol set, and N_{AM} represents the number of amplitude levels.

Information is embedded via phase rotation under the constraint:

$$\mathbf{w}_n^H\mathbf{a}(\theta_c) = d(\theta_c)e^{j\Omega_n}, \quad n = 1, 2, \ldots, N_{\text{PSK}}$$

where Ω_n is the PSK phase symbol, $d(\theta_c)$ is the desired gain in the communication direction, and N_{PSK} denotes the number of phase states.

Therefore, information can be embedded in the radar signals by amplitude modulation of the radar emission gain in the sidelobe region or by phase modulation of the entire spatial domain emission pattern. This beamforming approach allows radar and communication functionalities to be achieved simultaneously by keeping the main radar beam unchanged while modulating the sidelobe levels to convey information. Let $\tilde{G}(\theta, \mathcal{S}(\tau))$ and $\mathcal{S}(\tau)$ represent the modulated beam pattern and the embedded communication symbol, respectively, during the τth radar pulse. To perform the radar and communication simultaneously, the modulated beam pattern has to satisfy

$$\begin{aligned} &\tilde{G}(\theta, \mathcal{S}(\tau)) = P|\tilde{g}(\theta, \mathcal{S}(\tau))|^2 \simeq G(\theta), \\ &\forall\theta \in \Theta, \quad \mathcal{S}(\tau) \in \mathbb{D}_{\text{com}}. \end{aligned} \tag{2.3}$$

where $\mathbb{D}_{\text{com}}$ represents the communication symbol constellation, $\tilde{g}(\theta, \mathcal{S}(\tau))$ represents the modulation transmission gain, while $\Theta = (\theta_0 - (1/2)\Theta_{bw}, \theta_0 + (1/2)\Theta_{bw})$, θ_0 is the angle at the center of the radar's main beam and Θ_{bw} is the width of the radar main beam. Equation (2.3) is the prerequisite for maintaining the radar's main beam constant throughout the beampattern modulation processing.

Reference [4] implemented communication information embedding through ASK using sidelobe control and waveform diversity. The authors in [6] developed a far-field radiated emission design based on PSK. Reference [7] adopted AM to flexibly adjust the array structure for continuous sidelobe level control.

Lessons Learned **:** In ISAC systems, **AM** offers intuitive implementation by mapping communication symbols to sidelobe levels (SLLs) with minimal mainbeam impact; however, it suffers from limited data rates due to sidelobe attenuation constraints and high BER in low-SNR environments. **PM** provides superior noise immunity (e.g., PSK achieves 1–2 orders of magnitude lower BER than AM), enables mainbeam information embedding, and supports dynamic channel adaptation through non-coherent schemes; yet it requires precise channel phase estimation for coherent operation and may distort the radar waveform's ambiguity function, degrading range-Doppler resolution. **ASK** facilitates parallel transmission of M symbols in MIMO configurations and reduces BER via dual-SLL designs; nevertheless, it suffers from reduced per-waveform power (P/M) in MIMO systems, necessitating beamforming compensation while demanding strict waveform orthogonality.

(2) Index Modulation ISAC Systems: Index modulation (IM) fundamentally differs from traditional modulation by conveying information not through signal parameters (amplitude, frequency, or phase), but through the *index positions* of communication resource units (e.g., antennas, subcarriers, time slots). This technique maps combinatorial states of resource activation patterns to information bits, replacing conventional constellation point mapping. Crucially, in sensing-centric waveform design, IM preserves the original radar waveform with negligible performance degradation. The main diversity implementations are:

- **Spatial Diversity**: Utilizes antenna position differences, generating indexes by activating specific antenna combinations. Spatial diversity is typically combined with frequency diversity.
- **Frequency Diversity**: Employs frequency bands or subcarriers, creating indexes through activated frequency points. By leveraging frequency-hopping radar and antenna selection, multiple carrier frequencies are selected from the frequency set $\mathcal{F} = \{f_1, f_2, \ldots, f_K\}$ during each transmit pulse and transmitted through the chosen antennas $\mathcal{A} = \{a_1, a_2, \ldots, a_M\}$. Distinct combinations of the "frequency-antenna" indices represent different data symbols.

The MAJoRCom system [9] employs frequency agility and antenna selection to establish a spatial-frequency joint indexing scheme for symbol transmission, but suffers from power dispersion due to full-antenna transmission; Ma et al. [10] overcome the degrees of freedom (DoF) bottleneck by integrating FMCW waveforms with partial antenna activation and binary phase modulation. Baxter's proposed frequency-hopping code selection (FHCS) [11] maps symbols to hopping pattern combinations, while hybrid modulation schemes (e.g., PSK+FHCS [12]) achieve communication rate multiplication through *dual-domain (phase-frequency) encoding*.

Lessons Learned **:** IM enhances spectral efficiency by conveying additional information through the activation states of specific resource units (e.g., subcarriers, antennas), achieving higher data rates within identical bandwidth compared to conventional modulation schemes in ISAC system. It reduces power consumption in low-power scenarios by activating minimal resources (e.g., sparse subcarriers). However, IM performance is highly sensitive to Channel State Information (CSI), requiring precise channel estimation—a significant challenge in dynamic and time-varying channels. In massive MIMO systems, joint optimization of activation patterns and power allocation becomes prohibitively complex, necessitating low-complexity algorithmic solutions.

2.1.2 Communication-Centric ISAC Systems

Communication-centric ISAC systems are built on existing communication frameworks with added radar sensing capabilities. Currently, OFDM and OTFS have been widely adopted to support the design of such ISAC systems. The book considers ISAC signals where the same signal waveform simultaneously achieves both communication and radar functionalities. It does not consider schemes such as subcarrier allocation in OFDM systems, where radar subcarriers modulate radar data and communication subcarriers modulate communication data. In OFDM-based ISAC, the communication function transmits data via modulation symbols on subcarriers to ensure communication performance. While the sensing function estimates target range and velocity by detecting time delays and Doppler shifts in the received echo signals, thereby realizing sensing capabilities.

(1) OFDM-Based ISAC Systems: Consider a typical ISAC scenario where a BS transmits traditional OFDM signals for communication with users while simultaneously utilizing the OFDM echo signals to sense targets. Define $x(t)$ as the transmitted baseband OFDM signal.

$$x(t) = \sum_{m=0}^{N_s-1} \sum_{n=0}^{N_c-1} d_{m,n} e^{j2\pi f_n t} \mathrm{rect}\left(\frac{t - mT_{\mathrm{OFDM}}}{T_{\mathrm{OFDM}}}\right), \tag{2.4}$$

where N_s and N_c are the number of OFDM symbols and subcarriers within a single frame, respectively, $d_{m,n}$ represents the mth OFDM symbols on the nth subcarrier, while f_n denotes the frequency of the nth subcarrier. The overall duration of an OFDM symbol is given by $T_{\mathrm{OFDM}} = T_s + T_g$, where T_s and T_g are the durations of an elementary symbol and the cyclic prefix, respectively. Moreover, $\mathrm{rect}(t/T)$ depicts a rectangular window with a duration T.

The received reflected echo signal $y(t)$ can be obtained as

$$\begin{aligned} y(t) = \sum_{m=0}^{N_s-1} e^{j2\pi f_D t} \cdot \sum_{n=0}^{N_c-1} \mu(m,n) d_{m,n} e^{j2\pi f_n (t-\tau)} \\ \times \mathrm{rect}\left(\frac{t - mT_s - \tau}{T_s}\right), \end{aligned} \tag{2.5}$$

where $\mu(m,n)$ denotes the echo channel at the mth OFDM symbols on the nth subcarrier, f_D and τ are the Doppler shift and delay, respectively.

OFDM is often combined with other modulation methods, such as Linear Frequency Modulation (LFM), phase coding, and spread spectrum techniques, to enhance the sensing performance of OFDM-based ISAC signals [1].

- **OFDM + LFM:** Superimposes LFM modulation on OFDM subcarriers. Enhances range resolution by increasing the time-bandwidth product ($TB = B \cdot T$), suitable for long-distance target detection.

$$x(t) = \sum_{m=0}^{N_s-1} \sum_{n=0}^{N_c-1} d_{m,n} e^{j2\pi f_n t} \cdot e^{j\pi k t^2} \text{rect}\left(\frac{t - mT_{\text{OFDM}}}{T_{\text{OFDM}}}\right), \tag{2.6}$$

- **OFDM + Phase Coding:** Utilizes phase coding (e.g., Gold sequences, chaotic sequences) to optimize the ambiguity function sidelobes and reduce PAPR.

$$x(t) = \sum_{m=0}^{N_s-1} \sum_{n=0}^{N_c-1} a_{nm} \cdot e^{j2\pi f_n t} \cdot \text{rect}\left(\frac{t - mT_p}{T_p}\right), \tag{2.7}$$

  where the phase-coded communication symbol is defined as $a_{m,n} = d_{m,n}$ $\exp(j\phi_{m,n})$, where $d_{m,n}$ is the original modulation symbol, and $\phi_{m,n}$ denotes the phase coding sequence. And T_p denotes the symbol width of phase coding, satisfying $T_p = T_{\text{OFDM}}/N_s$.
- **OFDM + Spread Spectrum:** Extends the spectrum using DSSS technology, enhancing anti-jamming capability while reducing PAPR.

$$x(t) = \sum_{m=0}^{N_s-1} \sum_{n=0}^{N_c-1} \sum_{p=0}^{P-1} d_{m,n} a_m(p) e^{j2\pi f_n t} \cdot \text{rect}\left(\frac{t - pt_c - mT_{\text{OFDM}}}{T_{\text{OFDM}}}\right), \tag{2.8}$$

  where $a_m(p)$ denotes the p-th spread spectrum chip on the m-th OFDM symbol, taking values of ± 1 (e.g., BPSK spreading). And t_c represents the duration of a single spread spectrum chip, satisfying $t_c = T_{\text{OFDM}}/P$, i.e., each OFDM symbol is spread into P chips.

The instantaneous frequency of a LFM signal increases over time, thereby expanding the signal's time-bandwidth product and enabling high-resolution long-range radar sensing [13]. The ISAC signal combining OFDM with LFM effectively enhances Doppler sensitivity and reduces velocity estimation error [14].Reference [21] proposes an in-band full-duplex ISAC system based on chirp-scrambled OFDM, which utilizes the chirp characteristics of echo signals to achieve centimeter-level super-resolution distance estimation while maintaining communication performance equivalent to conventional OFDM. Reference [22] reveals the mathematical relationship between OFDM and FMCW waveforms, and accordingly proposes a novel method that fuses OFDM and FMCW through DFT matrix integration.

Reference [15] proposed a phase-coded OFDM signal design scheme for single point-scatterer targets. This OFDM signal comprises N subcarriers, with the bit stream mapped onto an M-bit sequence. Zhao et al. [16] designed a chaotic sequence-based phase-coded OFDM signal with large time-bandwidth product. They developed a method to extract well-correlated phase coding sequences from chaotic sequences, and further introduced subcarrier weighting to reduce the PAPR. Qi et al. [17] proposed an ISAC signal combining OFDM with phase coding using complete complementary codes, achieving high data information rate and precise target detection.

To enhance ISAC signal resolution, [18] proposed an OFDM signal combining spread spectrum with LFM. This design ensures signal orthogonality at the receiver while maintaining large time-bandwidth product. Beyond Gold sequences, references [19, 20] also developed ISAC signals using MSK and DSSS techniques.

***Lessons Learned*:** OFDM-LFM based ISAC systems deliver high-resolution sensing capabilities but suffer from high PAPR. OFDM with phase coding based ISAC systems achieve low PAPR and strong anti-jamming capability, while exhibiting sensitivity to carrier frequency offset and requiring complex synchronization algorithms.Spread Spectrum OFDM based ISAC systems provide robust anti-jamming performance and low PAPR characteristics at the cost of reduced bandwidth efficiency and high implementation complexity.

(2) OTFS-Based ISAC Systems: Although the OFDM has been widely adopted for ISAC system design, its performance degrades significantly in high-mobility scenarios due to severe Doppler shifts. To address this, OTFS modulation has been introduced for ISAC systems, offering enhanced performance in high-speed environments [48]. Unlike OFDM, which modulates data in the time-frequency domain, OTFS modulates data in the delay-Doppler (DD) domain. This allows for perfect alignment of time delays and Doppler shifts, preserving subcarrier orthogonality in high-mobility ISAC scenarios.

To transmit a block of symbols $\{x_{k,j}\}$, the inverse symplectic finite Fourier transform (ISFFT) is first applied to convert the data symbols into a block of samples $\{X[n,m]\}$ in the DD domain. This transformation can be expressed as

$$X[n,m] = \sum_{k=0}^{N-1}\sum_{l=0}^{M-1} x_{k,l} e^{j2\pi\left(\frac{nk}{N}-\frac{ml}{M}\right)}, \quad \text{for } n = 0,\ldots,N-1 \text{ and } m = 0,\ldots,M-1. \tag{2.9}$$

It generates the continuous-time signal based on the transformed samples, which can be expressed as

$$s(t) = \sum_{n=0}^{N-1}\sum_{n=0}^{M-1} X[n,m] g_{\text{tx}}(t-nT) e^{j2\pi m\Delta f(t-nT)}, \tag{2.10}$$

where $g_{tx}(t)$ is the transmit pulse shaping function and T is the symbol duration.

After transmission through a noiseless time-frequency selective channel, the received signal $r(t)$ can be expressed as

$$r(t) = \int h(t,\tau)s(t-\tau)d\tau = \sum_{p=0}^{P-1} h_p s(t-\tau_p)e^{j2\pi\nu_p t}. \quad (2.11)$$

By applying the symplectic finite Fourier transform (SFFT) to the received signal, the signal is transformed into the DD domain. In the DD domain, maximum likelihood estimation (MLE) can be employed to accurately estimate both the target's range and velocity.

In 2020, the effectiveness of OTFS waveforms for ISAC transmission was validated in [23], where the authors demonstrated that OTFS signals could achieve radar sensing with a bounded estimation error, without degrading communication performance. This work was later extended to the case of MIMO transmissions in [24], where a hybrid analog-digital (HAD) beamformer was designed for ISAC. Furthermore, Yuan et al. [25] investigated ISAC-assisted OTFS transmission in vehicular networks, proposing an effective beamforming prediction scheme. In [26], an OTFS-based ISAC framework was introduced, along with the concept of "spatial spreading", which facilitates the exploitation of channel characteristics in the DD-angle (DDA) domain.

Lessons Learned **:** Compared with OFDM, OTFS requires shorter CP while enabling longer sensing ranges and faster target tracking rates. Additionally, OTFS eliminates Inter-Carrier Interference (ICI), facilitating high-Doppler shift estimation and high-velocity target detection [27]. However, OTFS-based ISAC systems suffer from high computational complexity.

2.1.3 Joint ISAC Systems

While communication-centric and radar-centric integrated waveform designs can achieve a certain level of joint sensing and communication, they do not allow for a flexible trade-off between the two functions. Using radar sensing (communication) signals for communication (or sensing) presents significant challenges. However, joint waveform design addresses this issue effectively. Unlike previous methods, joint waveform design jointly optimizes waveform parameters based on performance metrics [28].

Reference [29] developed a joint waveform and filter design for coexisting MIMO radar and communication systems in clutter environments, maximizing radar SINR under communication rate constraints. Reference [30] achieved optimized target localization by minimizing the Cramér-Rao Bound (CRB) for Direction of Arrival (DOA) estimation subject to communication QoS constraints. Reference [31] implemented dual-functional cooperation through weighted-sum maximization of communication Mutual Information (MI) and sensing MI, incorporating spatio-temporal power optimization with training overhead. Reference [32] proposed a beamforming framework that matches the radar's desired beampattern while nulling interference to communication users in MU-MIMO systems. Reference [33] achieves superior symbol error rate performance and enhanced waveform design freedom through optimized waveform design that exploits rather than suppresses interference. Reference [34] establishes efficient information-theoretic performance in noisy environments by optimizing the Pareto boundary between Mutual Information and Minimum Mean-Square Error (MMSE). Reference [35] enhances sensing performance (sidelobe suppression and accuracy improvement) through optimized OFDM resource allocation (time, frequency, energy) and interpolation techniques using Schatten p-approximation. Reference [37] balances communication data rate and sensing accuracy through optimized cross-domain waveform strategies. Reference [38] derives Pareto boundary trade-offs between communication and sensing by optimizing channel capacity under sensing metrics with covariance mismatch constraints.

Lessons Learned **:** Joint ISAC waveform design achieves balanced dual-functional performance through multi-objective weighted optimization. However, most waveform optimization problems are non-convex, requiring approximate methods like Semi-Definite Relaxation (SDR) and Alternating Optimization (AO), where computational complexity grows exponentially with dimensionality.

2.2 ISAC Channel Models

To improve the communication and sensing performance of ISAC systems, realistic wireless channel models must be developed. An ISAC channel typically consists of both communication and sensing channels, as shown in Fig. 2.2. The sensing channel model can be built upon existing communication channel models by introducing elements relevant to sensing systems, such as the radar cross-section (RCS) and

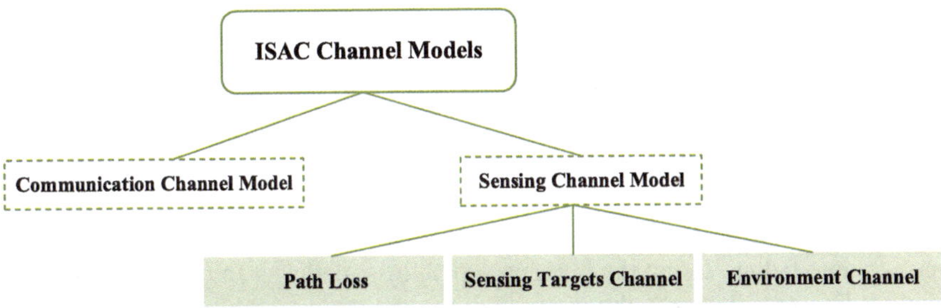

Fig. 2.2 ISAC channel models

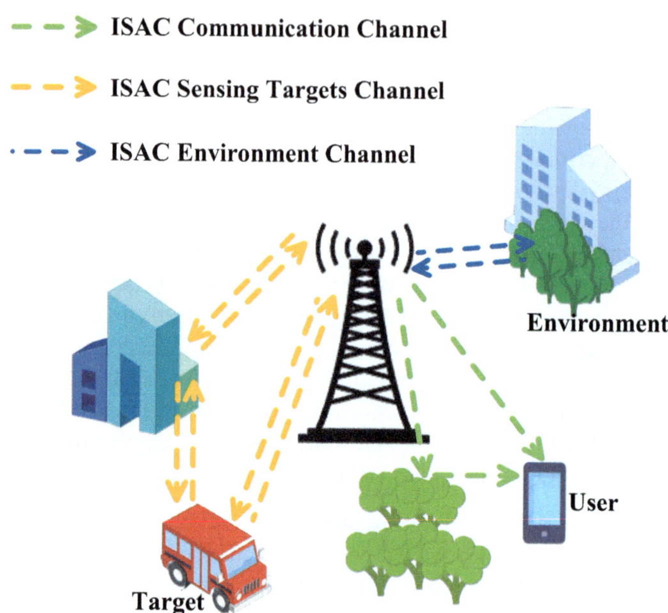

Fig. 2.3 Downlink ISAC in LOS and NLOS scenarios

other related features. In the following, we will investigate the communication and sensing channel models for downlink ISAC, encompassing both Line-of-Sight (LoS) and Non-Line-of-Sight (NLoS) scenarios, as depicted in Fig. 2.3.

2.2.1 Communication Channel Model

In ISAC systems, the communication channel modeling can follow the standard 3GPP channel models, which account for both large-scale and small-scale fading effects, such that

$$H(t) = \sqrt{PL} \cdot H_{u,s,n}(t), \tag{2.12}$$

where PL represents the large-scale fading, while $H_{u,s,n}(t)$ denotes the small-scale channel response.

Large-scale fading in communication channels primarily manifests in path loss and shadow fading. Using the free-space path loss model as an example, the large-scale fading model is given by

$$PL = 32.4 + 20\log_{10}(d_{\mathrm{3D}}) + 20\log_{10}(f_c) + X_\sigma \tag{2.13}$$

where shadow fading X_σ follows a Gaussian distribution with zero mean and σ^2 covariance. The path loss for other scenarios can be found in the 3GPP standards.

Small-scale fading in communication channels is primarily caused by multipath effects, where the signal travels through multiple paths with different delays and attenuations. Commonly, parameters such as delay spread, angle spread, and Doppler shift are used to describe the characteristics of small-scale fading in the channel. These parameters help capture the variations in signal strength due to the constructive and destructive interference of signals arriving via different paths, and are critical for accurate modeling and performance analysis in wireless communications.

In the presence of both LoS and NLoS paths, the small-scale channel response $H_{u,s,n}(t)$ can be modeled by Rician fading, which can be expressed as

$$
\begin{aligned}
H_{u,s,n}(t) =& \sqrt{\frac{1}{K+1}} H_{u,s,n}^{\mathrm{NLOS}}(t) + \delta(n-1)\sqrt{\frac{K}{K+1}} \times \begin{bmatrix} F_{rx,u,\theta}(\theta_{\mathrm{LOS,ZOA}}, \phi_{\mathrm{LOS,AOA}}) \\ F_{rx,u,\phi}(\theta_{\mathrm{LOS,ZOA}}, \phi_{\mathrm{LOS,AOA}}) \end{bmatrix}^T \\
&\times \begin{bmatrix} \exp(j\Phi_{\mathrm{LOS}}) & 0 \\ 0 & -\exp(j\Phi_{\mathrm{LOS}}) \end{bmatrix} \times \begin{bmatrix} F_{x,z,\theta}(\theta_{\mathrm{LOS,ZOD}}, \phi_{\mathrm{LOS,AOD}}) \\ F_{x,z,\phi}(\theta_{\mathrm{LOS,ZOD}}, \phi_{\mathrm{LOS,AOD}}) \end{bmatrix} \\
&\times \exp\left[j2\pi\lambda_0^{-1}(\bar{\Omega}_{rx,\mathrm{LOS}}^T \cdot \bar{d}_{rx,u}) + j2\pi\lambda_0^{-1}(\bar{\Omega}_{tx,\mathrm{LOS}}^T \cdot \bar{d}_{tx,s}) + j2\pi\nu_{\mathrm{LOS}} t \right].
\end{aligned} \tag{2.14}
$$

where K is the Rician factor and $H_{u,s,n}^{\mathrm{NLoS}}(t)$ represents the channel coefficient for the NLoS path at time t, which can be expressed as

$$
\begin{aligned}
H_{u,s,n}^{\mathrm{NLOS}}(t) =& \sqrt{\frac{P_n}{M}} \sum_{m=1}^{M} \begin{bmatrix} F_{m,u,\theta}(\theta_{n,m,\mathrm{ZOA}}, \phi_{n,m,\mathrm{AOA}}) \\ F_{m,u,\phi}(\theta_{n,m,\mathrm{ZOA}}, \phi_{n,m,\mathrm{AOA}}) \end{bmatrix}^T \times \begin{bmatrix} F_{x,z,\theta}(\theta_{n,m,\mathrm{ZOD}}, \phi_{n,m,\mathrm{AOD}}) \\ F_{x,z,\phi}(\theta_{n,m,\mathrm{ZOD}}, \phi_{n,m,\mathrm{AOD}}) \end{bmatrix} \\
&\times \begin{bmatrix} \exp(j\Phi_{n,m}^{\theta\theta}) & \sqrt{\kappa_{n,m}^{-1}} \exp(j\Phi_{n,m}^{\theta\phi}) \\ \sqrt{\kappa_{n,m}^{-1}} \exp(j\Phi_{n,m}^{\phi\theta}) & \exp(j\Phi_{n,m}^{\phi\phi}) \end{bmatrix} \times \exp\left[j2\pi\lambda_0^{-1}(\bar{\Omega}_{rx,n,m}^T \cdot \bar{d}_{rx,u}) \right. \\
&\left. + j2\pi\lambda_0^{-1}(\bar{\Omega}_{tx,n,m}^T \cdot \bar{d}_{tx,s}) + j2\pi\nu_{n,m} t \right] \delta(\tau - \tau_n).
\end{aligned} \tag{2.15}
$$

Note that the notations used in (2.14) and (2.15) are summarized in Table 2.1.

2.2.2 Sensing Channel Model

In ISAC systems, the sensing component relies on detecting and analyzing the reflected signals from targets in the environment. To achieve accurate target detection and sensing, it is necessary to model a two-way multipath sensing channel, which consists of two key links:

- The link from the signal transmitter to the sensing target.

Table 2.1 Parameter definitions for communication in ISAC systems

Parameter	Description
$F_{rx,u,\theta}$ $F_{rx,u,\phi}$	Field pattern of receive antenna element u in spherical coordinates (θ, ϕ)
$F_{tx,s,\theta}$ $F_{tx,s,\phi}$	Field pattern of transmit antenna element s in spherical coordinates (θ, ϕ)
$\phi_{\text{LOS,AOD}}$ $\theta_{\text{LOS,ZOD}}$	LOS azimuth angle between base station and communication receiver
$\phi_{\text{LOS,AOA}}$ $\theta_{\text{LOS,ZOA}}$	LOS elevation angle between base station and communication receiver
$\bar{d}_{tx,s}$	Transmission vector from the transmitter to the receiver
$\bar{d}_{rx,u}$	Reception vector from the transmitter to the receiver
P_n	Transmit power of the sender
M	Number of sub-paths
τ_n	Generated cluster delay
$\bar{\Omega}^T_{tx,n,m}$ $\bar{\Omega}^T_{rx,n,m}$	Unit vector of departure and arrival angles
$v_{n,m}$	Doppler frequency factor

- The link from the sensing target back to the receiver.

The sensing channel is thus formed by both the target-specific and environmental characteristics, which can be expressed as

$$H(\tau, t) = \sum_{k=1}^{T} SL_k PL_k H^k_{u,s}(\tau, t) + SL_e PL_e H^r_{u,s}(\tau, t), \tag{2.16}$$

where T represents the number of sensing targets, $H^k_{u,s}(\tau, t)$ denotes the sensing channel coefficient for the kth target, and $H^e_{u,s}(\tau, t)$ corresponds to the channel coefficient for the environment. Additionally, PL_k and SL_k represent the path loss and shadow fading for the kth target, respectively, while PL_e and SL_e indicate the path loss and shadow fading for the environment.

2.2.2.1 Path Loss in Sensing Channel

The path loss calculation formula for the ISAC system can be modified based on the existing path loss calculations defined by 3GPP. The current 3GPP path loss formula

Table 2.2 Typical RCS values of reflective objects at microwave frequencies

Reflective Object	RCS (m^2)	RCS (dBsm)
Helicopter	3	4.8
Small vessel/small UAV	0.02	−17
Small boat (20-30 ft)	3	4.8
Car	100	20
Truck/Van	200	23
Bicycle	2	3
Human	1	0
Large bird	10^{-2}	−20
Medium bird	10^{-3}	−30
Large insect	10^{-4}	−40
Small insect	10^{-5}	−50

Note The two RCS values are related by the equation:$\mathrm{RCS}(dBsm) = 10\log_{10}(\mathrm{RCS}(m^2))$

describes a unidirectional link from the transmitter to the receiver, incorporating the effects of the receiver's antenna aperture, while the sensing target only reflects the signal.

If the path loss from the sensing transmitter to the sensing target is denoted as $PL(d_1)$ and the path loss from the sensing target to the sensing receiver as $PL(d_2)$, applying the 3GPP TR 38.901 path loss model for both paths effectively calculates the receiver's antenna aperture effect twice, without accounting for the influence of scattering from the sensing target. To rectify this, it is necessary to subtract the antenna aperture effect once and incorporate the effect of the target's RCS. Therefore, the path loss calculation for the ISAC system, $PL_s(d_1, d_2)$, can be expressed as

$$PL_s(d_1, d_2) = PL(d_1) + PL(d_2) + 10\log_{10}\left(\frac{\lambda^2}{4\pi}\right) - 10\log_{10}(\sigma_{\mathrm{RCS}}), \quad (2.17)$$

where $PL(d_1)$ and $PL(d_2)$ can be calculated using the existing path loss formulas defined in 3GPP, or alternatively, the free-space path loss formula. Here, λ represents the wavelength in meters (m), and σ_{RCS} denotes the RCS of the sensing target in square meters (m^2). Specifically, the calculation methods for $PL(d_1)$ and $PL(d_2)$ should refer to the modeling approaches used for the communication channel.

Assuming the sensing target is a point target, we consider the RCS to be constant regardless of the direction of incidence or reflection, thus using a specific value for RCS. Different sensing targets typically have varying RCS values. Table. 2.2 presents typical RCS values for various reflectors in the microwave frequency range.

2.2.2.2 Sensing Targets Channel

The sensing channel for the targets utilizes the method outlined in 3GPP TR 38.901 to generate the sensing channel coefficients. However, it is essential to incorporate additional factors such as the absolute delay of the clusters and the Doppler effects caused by the relative velocities of the targets with respect to both the sensing transmitter and receiver. In scenarios where both LoS and NLoS paths exist, the model can be constructed to represent a single LoS echo path from the sensing transmitter to the target, combined with multiple NLoS echo paths. This approach effectively captures the propagation characteristics by merging the LoS path with the NLoS paths using the Rician factor K_R, allowing for a comprehensive representation of the LoS echo as part of the channel model, such that

$$H_{u,s}^{\mathrm{LOS}}(\tau,t)=\sqrt{\frac{1}{K_R+1}}H_{u,s}^{\mathrm{NLOS}}(\tau,t)+\sqrt{\frac{K_R}{K_R+1}}H_{u,s,1}^{\mathrm{LOS}}(t)\delta(\tau-\tau_1-d_{3D}^{e}/c), \tag{2.18}$$

where the channel coefficient $H_{u,s,1}^{\mathrm{LOS}}(t)$ and $H_{u,s}^{\mathrm{NLOS}}(\tau,t)$ represent the LoS and NLoS echo path, respectively.

The LoS echo channel coefficient $H_{u,s,1}^{\mathrm{LOS}}(t)$ can be expressed as

$$\begin{aligned} H_{u,s,1}^{\mathrm{LOS}}(t)=&\begin{bmatrix} F_{rx,u,\theta}(\theta_{\mathrm{LOS,ZOA}},\phi_{\mathrm{LOS,AOA}}) \\ F_{rx,u,\phi}(\theta_{\mathrm{LOS,ZOA}},\phi_{\mathrm{LOS,AOA}}) \end{bmatrix}^T \begin{bmatrix} 1 & 0 \\ 0 & -1 \end{bmatrix} \\ &\times \begin{bmatrix} F_{tx,s,\theta}(\theta_{\mathrm{LOS,ZOD}},\phi_{\mathrm{LOS,AOD}}) \\ F_{tx,s,\phi}(\theta_{\mathrm{LOS,ZOD}},\phi_{\mathrm{LOS,AOD}}) \end{bmatrix} \exp\left(-j2\pi\frac{d_{3D}}{\lambda_0}\right) \\ &\times \exp\left(j2\pi\frac{\hat{r}_{rx,\mathrm{LOS}}^T\cdot \bar{d}_{rx,u}}{\lambda_0}\right)\exp\left(j2\pi\frac{\hat{r}_{tx,\mathrm{LOS}}^T\cdot \bar{d}_{tx,s}}{\lambda_0}\right) \\ &\times \exp\left(j2\pi\frac{\hat{r}_{rx,\mathrm{LOS}}^T\cdot \bar{v}_r}{\lambda_0}t\right)\exp\left(j2\pi\frac{\hat{r}_{tx,\mathrm{LOS}}^T\cdot \bar{v}_t}{\lambda_0}t\right). \end{aligned} \tag{2.19}$$

The random phase associated with this LoS path can be considered a reference benchmark, effectively serving as the normalized phase. This normalization allows for clearer interpretation of the phase shifts caused by environmental factors and target movements, facilitating more accurate estimation of the channel characteristics in the sensing system. In (2.19), the second term utilizes the deterministic phase expression for the LoS path as defined in 3GPP TR 38.901. However, in certain scenarios where the LoS echo reflects off the sensing target, the phase may undergo random changes.

Moreover, the NLoS echo channel coefficient $H_{u,s}^{\mathrm{NLOS}}(\tau,t)$ can be expressed as

$$H_{u,s}^{\text{NLOS}}(\tau,t)=\sum_{i=1}^{3}\sum_{m\in R_i}H_{u,s,m}^{\text{NLOS}}(t)\delta\Big(\tau-\tau_{1j}-d_{3D}/c\Big),\tag{2.20}$$

where $H_{u,s,m}^{\text{NLOS}}(t)$ represents the channel coefficient for the NLoS echo paths in the cluster of LoS echoes from the sensing transmitter to the sensing target and then to the sensing receiver, which can further expressed as

$$\begin{aligned}H_{u,s,m}^{\text{NLOS}}(t)=&\sqrt{\frac{P_1}{M}}\begin{bmatrix}F_{rx,u,\theta}(\theta_{m,\text{ZOA}},\phi_{m,\text{AOA}})\\F_{rx,u,\phi}(\theta_{m,\text{ZOA}},\phi_{m,\text{AOA}})\end{bmatrix}^T\times\begin{bmatrix}F_{tx,s,\theta}(\theta_{m,\text{ZOD}},\phi_{m,\text{AOD}})\\F_{tx,s,\phi}(\theta_{m,\text{ZOD}},\phi_{m,\text{AOD}})\end{bmatrix}\\&\times\begin{bmatrix}\exp(j\Phi_m^{\theta\theta}) & \sqrt{\kappa_m^{-1}}\exp(j\Phi_m^{\theta\phi})\\\sqrt{\kappa_m^{-1}}\exp(j\Phi_m^{\phi\theta}) & \exp(j\Phi_m^{\phi\phi})\end{bmatrix}\\&\times\exp\left(j2\pi\frac{\hat{r}_{rx,m}^T\cdot\bar{d}_{rx,u}}{\lambda_0}\right)\exp\left(j2\pi\frac{\hat{r}_{tx,m}^T\cdot\bar{d}_{tx,s}}{\lambda_0}\right)\\&\times\exp\left(j2\pi\frac{\hat{r}_{rx,m}^T\cdot\bar{v}_r}{\lambda_0}t\right)\exp\left(j2\pi\frac{\hat{r}_{tx,m}^T\cdot\bar{v}_t}{\lambda_0}t\right).\end{aligned}\tag{2.21}$$

Note that the notations used in from (2.18) to (2.21) are summarized in Table 2.3.

Considering the channel coefficients for multiple sensing targets, the above steps are applied to each target, combining large-scale fading to form the overall sensing channel for the targets.

2.2.2.3 Environment Channel

The environmental channel is generated independently using statistical methods, making it uncorrelated with the sensing targets. This can be viewed as a statistical model of the environmental targets, where the NLoS clusters correspond to statistically defined environmental targets. The steps from 3GPP TR 38.901 are utilized to generate the environmental channel parameters.

The channel coefficients for the N NLoS clusters between the sensing transmitter and receiver, denoted as $H_{u,s,n,m}^{\text{NLOS}}(\tau,t)$, and the channel coefficient for the LoS cluster, $H_{u,s,1}^{\text{LOS}}(t)$, are generated based on the methods outlined in 3GPP TR 38.901. The LoS channel coefficient for the environment can be expressed as

$$H_{u,s}^{\text{LOS}}(\tau,t)=\sqrt{\frac{1}{K_R+1}}H_{u,s}^{\text{NLOS}}(\tau,t)+\sqrt{\frac{K_R}{K_R+1}}H_{u,s,1}^{\text{LOS}}(t)\delta(\tau-\tau_1-d_{3D}^e/c).\tag{2.22}$$

Furthermore, the NLoS channel coefficient for the environment is given by

Table 2.3 Parameter definitions for sensing in ISAC systems

Parameter	Description
$F_{rx,u,\theta}$ $F_{rx,u,\phi}$	Field pattern of receive antenna element u in spherical coordinates (θ, ϕ)
$F_{tx,s,\theta}$ $F_{tx,s,\phi}$	Field pattern of transmit antenna element s in spherical coordinates (θ, ϕ)
$\phi_{\text{LOS,AOD}}$ $\theta_{\text{LOS,ZOD}}$	LOS azimuth angle between base station and sensing target
$\phi_{\text{LOS,AOA}}$ $\theta_{\text{LOS,ZOA}}$	LOS elevation angle between sensing target and communication receiver
$\hat{r}_{tx}^{T}$	Transmission vector from sensing transmitter to target
$\hat{r}_{rx}^{T}$	Reception vector from sensing target to the receiver
P_1	Sensing transmit power for the target
M	Number of sub-paths for LOS echo cluster of each target
$\tau_{1,i}$	Generated cluster delay
$\bar{v}_r$	Velocity of the sensing receiver relative to the target
d_{3D}	Distance between the sensing sender to the sensing target to the sensing receiver

$$\begin{aligned} H_{u,s}^{\text{NLOS}}(\tau, t) = &\sum_{n=1}^{2}\sum_{i=1}^{3}\sum_{m\in R_i} H_{u,s,n,m}^{\text{NLOS}}(t)\delta\Big(\tau - \tau_{n,i} - d_{3D}^{e}/c\Big) \\ &+ \sum_{n=3}^{N} H_{n,s,n}^{\text{NLOS}}(t)\delta(\tau - \tau_n - d_{3D}^{c}/c). \end{aligned} \tag{2.23}$$

The large-scale path loss includes the path loss between the sensing transmitter and the sensing target, as well as between the sensing target and the sensing receiver. It also accounts for the fading caused by signal reflection and scattering from the sensing target, represented through RCS modeling.

2.3 Conclusion

In this chapter, we provided an overview of the basic principles of ISAC, focusing on waveform design and channel model. In terms of waveform design, three representative waveform principles were introduced:radar-centric, communication-centric, and joint design. In terms of channel modeling, we explored communication and sensing channel models, emphasizing path loss and small-scale fading for the communication channel model; and distinguishing between target echo channel and environment channel for the sensing channel model. These insights lay the foundation for FD-ISAC, which will be discussed in subsequent chapters.

References

1. Wei Z, Qu H, Wang Y et al (2023) Integrated sensing and communication signals toward 5G-A and 6G: a survey. IEEE Internet Things J 10(13):11068–11092
2. Chiriyath AR, Paul B, Bliss DW (2017) Radar-communications convergence: coexistence, cooperation, and co-design. IEEE Trans Cognit Commun Netw 3(1):1–12
3. Hassanien A, Amin MG, Zhang YD, Ahmad F (2016) Phase-modulation based dual-function radar-communications. IET Radar Sonar Navigation 10:1411–1421
4. Hassanien A, Amin MG, Zhang YD, Ahmad F (2016) Dual-function radar-communications: information embedding using sidelobe control and waveform diversity. IEEE Trans Signal Process 64:2168–2181
5. de Oliveira Ferreira A, Sampaio-Neto R, Fortes J M (2018) Robust sidelobe amplitude and phase modulation for dual-function radar-communications. Trans Emerg Telecommun Technol 29(9)
6. McCormick P M, Blunt S D, Metcalf J G (2017) Simultaneous radar and communications emissions from a common aperture, part I: Theory. In: Proceedings of IEEE radar conference, pp 1685–1690
7. Feng L, Cui G, Kong L, Zhang X (2019) A RadCom system with flexible array controls. In: Proceedings of IEEE radar conference, pp 1–5
8. Yang J, Cui G, Yu X, Kong L (2020) Dual-use signal design for radar and communication via ambiguity function sidelobe control. IEEE Trans Veh Technol 69(9):9781–9794
9. Huang T, Shlezinger N, Xu X, Liu Y, Eldar YC (2020) MAJoRCom: a dual-function radar communication system using index modulation. IEEE Trans Signal Process 68:3423–3438
10. Ma D, Shlezinger N, Huang T, Liu Y, Eldar YC (2021) FRaC: FMCW-based joint radar-communications system via index modulation. IEEE J Sel Topics Signal Process 15(6):1348–1364
11. Baxter W, Aboutanios E, Hassanien A (2018) Dual-function MIMO radar-communications via frequency-hopping code selection. In: Proceedings of 52nd asilomar conference signals system computing, pp 1126–1130
12. Wu K, Zhang JA, Huang X, Guo YJ, Heath RW (2021) Waveform design and accurate channel estimation for frequency-hopping MIMO radar-based communications. IEEE Trans Commun 69(2):1244–1258
13. Wang W-Q (2012) An orthogonal frequency division multiplexing radar waveform with a large time-bandwidth product. Defence Sci J 62(6):427–430
14. Li M, Wang W-Q, Zheng Z (2018) Communication-embedded OFDM chirp waveform for delay-Doppler radar. IET Radar Sonar Navig 12(3):353–360
15. Tian X, Zhang T, Zhang Q, Song Z (2017) Waveform design and processing in OFDM based radar-communication integrated systems. In: Proceedings of IEEE/CIC international conference on communication China (ICCC), pp 1–6
16. Zhao J, Huo K, Li X (2014) A chaos-based phase-coded OFDM signal for joint radar-communication systems. In: Proceedings of 12th international conference on signal processing (ICSP), pp 1997–2002
17. Qi L, Yao Y, Huang B, Wu G (2019) A phase-coded OFDM signal for radar-communication integration. In: Proceedings of IEEE international symposium phased array system and technology (PAST), pp 1–4
18. Cheng S-J, Wang W-Q, Shao H-Z (2015) Spread spectrum-coded OFDM chirp waveform diversity design. IEEE Sensors J 15(10):5694–5700

19. Liang T, Li Z, Wang M, Fang X (2019) Design of radar-communication integrated signal based on OFDM. In: Proceedings of international conference on artificial intelligence communication network, pp 108–119
20. Tang L, Zhang K, Dai H, Zhu P, Liang Y-C (2019) Analysis and optimization of ambiguity function in radar-communication integrated systems using MPSK-DSSS. IEEE Wireless Commun Lett 8(6):1546–1549
21. Cho W, Chang K, Shin W, Kim Y, Ko Y-J (2025) OFDM-based in-band full-duplex ISAC systems. IEEE Wireless Commun Lett 14(2):365–369
22. Bouziane A, Eddine Zegrar S, Arslan H (2025) A novel OFDM-FMCW waveform for low-complexity joint sensing and communication. IEEE Wirel Commun Lett 14(2):425–429
23. Gaudio L, Kobayashi M, Caire G, Colavolpe G (2020) On the effectiveness of OTFS for joint radar parameter estimation and communication. IEEE Trans Wirel Commun 19(9):5951–5965
24. Gaudio L, Kobayashi M, Caire G, et al. (2020) Hybrid digital-analog beamforming and MIMO radar with OTFS modulation. arXiv:2009.08785
25. Yuan W, Wei Z, Li S et al (2021) Integrated sensing and communication-assisted orthogonal time frequency space transmission for vehicular networks. IEEE J Sel Top Signal Process 15:1515–1528
26. Li S, Yuan W, Liu C et al (2022) A novel ISAC transmission framework based on spatially-spread orthogonal time frequency space modulation. IEEE J Sel Areas Commun 40(6):1854–1872
27. Raviteja P, Phan K T, Hong Y, Viterbo E (2019) Orthogonal time frequency space (OTFS) modulation based radar system. In: Proceedings of IEEE radar conference (RadarConf), pp 1–6
28. Zhou W, Zhang R, Chen G, Wu W (2022) Integrated sensing and communication waveform design: a survey. IEEE Open J Commun Soc 3:1930–1949
29. Li B, Petropulu AP (2017) Joint transmit designs for coexistence of MIMO wireless communications and sparse sensing radars in clutter. IEEE Trans Aerosp Electron Syst 53(6):2846–2864
30. Cheng Z, Liao B, Shi S, He Z, Li J (2019) Co-design for overlaid MIMO radar and downlink MISO communication systems via Cramér-Rao bound minimization. IEEE Trans Signal Process 67(24):6227–6240
31. Yuan X et al (2021) Spatio-temporal power optimization for MIMO joint communication and radio sensing systems with training overhead. IEEE Trans Veh Technol 70(1):514–528
32. Liu F, Masouros C, Li A, Sun H, Hanzo L (2018) MU-MIMO communications with MIMO radar: from co-existence to joint transmission. IEEE Trans Wirel Commun 17(4):2755–2770
33. Liu T, Guo Y, Lu L, Xia B (2022) Waveform design for integrated sensing and communication systems based on interference exploitation. J Commun Inf Netw 7(4):447–456
34. Wang S, Chen L, Zhou J, Chen Y, Han K, You C (2024) Unified ISAC pareto boundary based on mutual information and minimum mean-square error estimation. IEEE Trans Commun 72(11):6783–6795
35. Mura S, Tagliaferri D, Mizmizi M, Spagnolini U, Petropulu A (2025) Optimized waveform design for OFDM-based ISAC systems under limited resource occupancy. IEEE Trans Wireless Commun 24(6):5241–5254
36. Wu K, Zhang JA, Ni Z, Huang X, Guo YJ, Chen S (2024) Joint communications and sensing employing optimized MIMO-OFDM signals. IEEE Internet Things J 11(6):10368–10383
37. Zhang F, Mao T, Liu R, Han Z, Chen S, Wang Z (2024) Cross-domain dual-functional OFDM waveform design for accurate sensing/positioning. IEEE J Sel Areas Commun 42(9):2259–2274
38. Liu Y, Huang T, Zheng Z, He B, Huangfu W, Wang X, Zhang H, Long K (2025) Analysis of pareto boundary in MIMO ISAC: from the perspective of instantaneous covariance mismatch. IEEE Trans Wireless Commun 24(1):555–570

Chapter 3
Key Technologies of FD

Abstract Full-Duplex technology, by enabling simultaneous transmission and reception in the same frequency band, not only enhances communication capacity but also eliminates the sensing blind spots inherent in conventional TDD. This dual advantage positions FD as a pivotal breakthrough for advancing ISAC system performance. This chapter introduces FD technology within the ISAC framework, focusing on the FD system model, FD self-interference (SI) signal model, and interference cancellation techniques. Initially, we overview the FD system model, exploring echo signal characteristics to ensure synchronous signal transmission and reception within the same frequency band. This synchronization supports the seamless integration of communication and sensing, laying a foundation for FD-ISAC systems. We then summarize methods for modeling FD-ISAC system self-interference, considering nonlinearities from wireless multipath channels and transceiver links. The model includes direct path leakage, reflected interference, and hardware-induced distortions, providing an accurate reflection of real-world interference characteristics and a basis for interference mitigation strategies. Finally, we detail self-interference cancellation techniques across spatial, analog, and digital domains. Spatially, advanced antenna designs and beamforming suppress interference. Analog solutions involve high-linearity amplifiers and optimized front-end designs to minimize leakage, while digital strategies use adaptive algorithms and machine learning to enhance cancellation precision. This multi-tiered approach ensures efficient interference management in complex environments, optimizing both communication and sensing functionalities in FD-ISAC systems.

3.1 FD System Model

Full-duplex technology occupies a pivotal role in modern wireless communication systems primarily due to its capability to simultaneously transmit and receive signals, significantly enhancing spectral efficiency and overall system performance. However, a major challenge arises from the reception of high-power signals emitted by the transmitter being captured by the receiving antenna of the same device, leading to strong self-interference. This self-interference often substantially exceeds

C. Du et al., *Full-Duplex Integrated Sensing and Communication Systems*,
https://doi.org/10.1007/978-981-92-0470-0_3

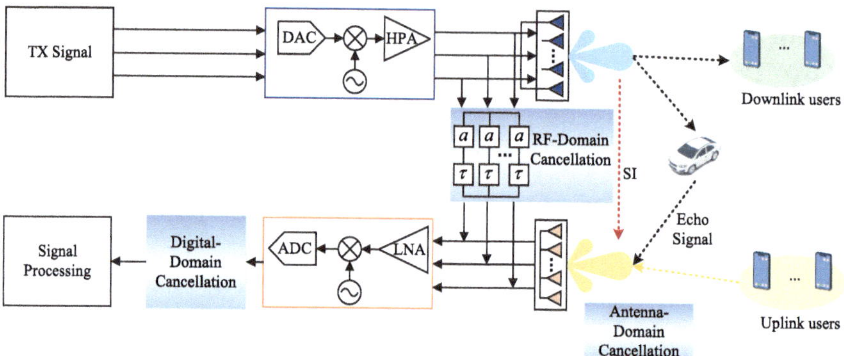

Fig. 3.1 A typical model of FD system

the power of useful received signals, severely degrading signal quality and the reliability of communication. Consequently, the crux of full-duplex technology lies in effectively mitigating self-interference. The design of full-duplex system models must, therefore, incorporate robust self-interference cancellation (SIC) mechanisms at both hardware architecture and algorithmic levels to ensure stable and efficient system operation.

Figure 3.1 illustrates a typical model of a full-duplex system, detailing the flow of signals along the transmission and reception paths and the strategies for self-interference suppression across various domains (antenna domain, RF domain, and digital domain). In full-duplex communication systems, the transmission and reception stages are crucial for ensuring simultaneous signal transfer and reception. The transmission process begins with baseband signal processing where the "TX Signal" represents the processed digital signal, which is then converted into an analog signal by a digital-to-analog converter (DAC). This analog signal is subsequently amplified by a high-power amplifier (HPA) to ensure that the emissions via the transmit antenna are sufficiently strong to radiate effectively outward. Not only does this radiated signal transmit data to downlink users, but it may also interact with objects in the surrounding environment (such as vehicles), generating echo signals.

At the receiving end, the receive antenna captures echo signals from the surroundings as well as communication signals from uplink users. Due to the inherent characteristics of full-duplex systems, the receive antenna also picks up signals transmitted by its own system, leading to self-interference. These signals are first amplified by a low-noise amplifier (LNA) and then converted into digital signals by an analog-to-digital converter (ADC). Initial self-interference mitigation is conducted through antenna and RF domain techniques, including optimized antenna design and adjustments at the RF frontend. In the digital domain, the signal processing module further processes these signals, employing digital domain techniques to alleviate the effects of self-interference and ensuring the clarity and reliability of received signals. Collectively, these measures ensure the effective reception and processing of signals, supporting the continuity and stability of full-duplex communications.

3.2 FD SI Signal Model

SI signal model is the basis for understanding and eliminating SI. An accurate model that characterizes the SI signal, including its amplitude, phase, and frequency characteristics, is essential for designing the appropriate cancellation algorithms. SI in FD-ISAC systems is mainly classified into linear and nonlinear components, and in order to adequately suppress SI, the SI signal model must be sufficiently accurate to characterize the variations that the transmitted signal undergoes in the transceiver, including the changes introduced by the wireless multipath propagation nonlinear components introduced by the channel and transceiver link, etc. The SI signal is modeled as follows:

3.2.1 Linear Signal Model

Linear SI is formulated as a non-causal linear function of the baseband transmit signal. This primarily takes into account the impact of the multipath channel, which encompasses all the paths from the transmitter to the receiver, including the dominant close-in LoS path as well as various reflected and scattered paths. The linear signal model for SI is represented as [1]

$$y_{\mathrm{LIN}}[n] = \sum_{m=0}^{M-1} h_m x[n-m] \tag{3.1}$$

where $x[n]$ is the baseband transmit signal, h_m is the unknown coefficient of the signal model, and M denotes the number of memory taps. However, due to the inherent limitations of linear models in fully characterizing the constituents of SI, it becomes imperative to extend the model to include nonlinear components.

Lessons Learned **:** The linear model offers low computational complexity yet fails to characterize nonlinear components, rendering it applicable only in idealized channel scenarios.

3.2.2 Nonlinear Signal Model

In full-duplex communication systems, the sources of nonlinear self-interference include multipath effects, power amplifiers, mixers, and low noise amplifiers, among others. Multipath effects occur when signals encounter obstacles such as buildings or terrain, leading to reflections, refractions, and scattering. This can cause changes in the phase and amplitude of signals at the receiver, introducing nonlinear distortion. Power amplifiers boost the power of transmission signals, but their nonlinear characteristics, such as saturation and compression, can produce harmonic distortion. Mixers, serving as frequency conversion devices, combine input signals with local oscillator signals to produce the desired frequency signal, but their suboptimal performance can lead to additional interference and distortion. Low noise amplifiers are used to amplify weak received signals; although designed to be low-noise, their nonlinear amplification at high gains can also introduce distortion. The interaction of these components increases the complexity of self-interference in full-duplex systems, posing greater challenges for modeling full-duplex nonlinear self-interference and designing interference cancellation systems.

Existing models of nonlinear self-interference in full-duplex systems are depicted as shown in Fig. 3.2, with different models having varying effects on suppressing types of nonlinear components. The following will provide a detailed introduction to each model type. In model construction, polynomial basis functions are primarily used to model nonlinear components, reconstructing and canceling self-interference signals based on the constructed models. Commonly, the approach involves establishing a model between the system's input and output signals, treating the entire communication system as a black-box system, necessitating the construction of complex basis functions. This approach requires a nuanced understanding of the system's dynamics and the specific nonlinearities involved to effectively mitigate self-interference and enhance system performance.

- Volterra series model:

The Volterra series model [2] is a type of functional series that can simulate the memory effects of signals as well as nonlinearities introduced by components. It is widely recognized as one of the most comprehensive models for representing nonlinear signals. The general mathematical formulation of the Volterra series model is expressed as follows:

$$y(n) = \sum_{d=1}^{D} \left(\sum_{l_1=0}^{L-1} \sum_{l_2=0}^{L-1} \cdots \sum_{l_d=0}^{L-1} h\left(l_1, l_2, \ldots, l_d\right) \prod_{i=1}^{d} x\left(n - l_i\right) \right) \tag{3.2}$$

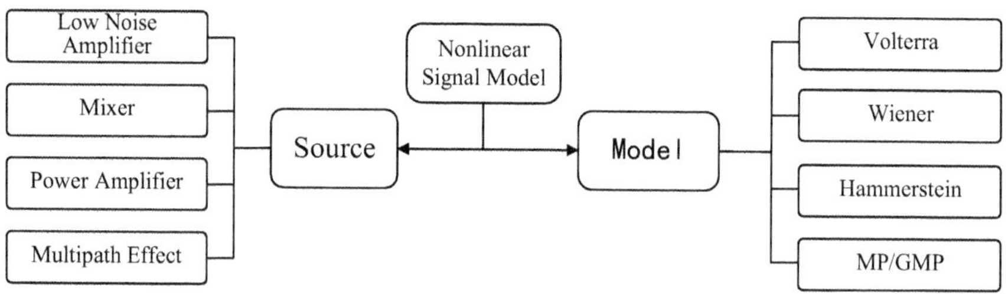

Fig. 3.2 Nonlinear signal model

In the Volterra series model, $x(n)$ represents the input signal to the system, and $y(n)$ denotes the nonlinear output. The coefficients h correspond to the weights associated with different basis functions, d indicates the order of the Volterra series terms being considered, and D represents the maximum nonlinearity order accounted for in the model. L signifies the memory depth, which reflects the input signal's memory delay.

This setup allows the model to account for how past and current values of the input signal contribute to the output, capturing both the immediate and delayed effects of the input on the output. The model's ability to incorporate multiple orders of nonlinearity and extensive memory makes it exceptionally powerful for modeling complex systems where interactions within the signal occur over extended periods. However, the increase in model complexity with higher orders and greater memory depths necessitates sophisticated computational strategies to manage the exponentially growing number of parameters.

The Volterra series model comprehensively incorporates memory characteristics and nonlinear features, accurately reflecting complex nonlinear dynamics and is widely recognized as a classical modeling method [2]. Particularly in the modeling of power amplifiers, the Volterra series model provides a solid theoretical foundation, facilitating its subsequent application in related devices [3]. For instance, the application of the Volterra series model in power amplifiers has been proven particularly effective in suppressing nonlinearity, effectively compensating for nonlinearities and IQ imbalances introduced by power amplifiers (PA) and low noise amplifiers (LNA). It has been reported that using the Volterra model for digital domain self-interference cancellation in a single antenna system can significantly enhance performance, improving cancellation capabilities by 20 dB compared to traditional linear methods [4]. Additionally, a study [5] utilized a Volterra signal model with sparse memory to model residual self-interference and employed the Recursive Least Squares (RLS) method for cancellation. This approach significantly reduces computational costs while simulating longer delays and other memory effects, demonstrating superior processing capabilities compared to other models. Despite the Volterra series model's excellent capability in nonlinear signal modeling, its high computational complexity remains a significant challenge in engineering implementation.

In practical applications, since even-order components of signals are easily filtered out, the Volterra series is typically represented as a combination of odd orders. While the Volterra series can effectively represent power amplifiers under the influence of memory effects and multi-order nonlinearity, the model's high complexity limits its practical engineering implementation. Therefore, it becomes necessary to appropriately modify the signal model based on the input signals, application scenarios, and channel conditions to reduce computational demands and enhance the feasibility of engineering implementation.

If only the diagonal terms are retained in the Volterra series and the influence of the orthogonal terms is ignored, the series can be expressed as a Hammerstein model. The Hammerstein model is a form of filter cascade that combines a filter representing memory effects with a power series model representing nonlinearity.

This model simplifies the expression of the Volterra series. The mathematical model is as follows.

This configuration effectively reduces the complexity of the full Volterra series by focusing on a single nonlinear function following a linear memory system, which makes the Hammerstein model particularly useful for applications where the nonlinearity is predominantly concentrated after a linear dynamic effect.

Reference [6] details an iterative method for identifying nonlinear systems from noisy input and output samples, tailored specifically for the Hammerstein model. This model incorporates a memoryless polynomial and a linear system, and iteratively adjusts the parameters of the linear system's transfer function and the polynomial's nonlinear coefficients to minimize mean square error. This approach has been effectively applied to estimate the channel responses of all nonlinear basis functions [7], successfully addressing nonlinear issues caused by IQ imbalance and power amplifier (PA) distortion. According to experimental results published in [8], this method achieved up to 46 dB of digital domain cancellation performance at a central frequency of 2.46 GHz and a bandwidth of 20 MHz. Furthermore, Ref. [9] proposed a Parallel Hammerstein (PH) structure for processing signals, which demonstrated superior cancellation effects under equivalent transmission power conditions. Additionally, the PH structure can independently address IQ imbalance and crosstalk issues in Multiple-Input Multiple-Output (MIMO) systems [10]. Despite the Hammerstein model's effectiveness in nonlinear self-interference cancellation, the multitude of highly correlated basis functions can potentially diminish the cancellation performance.

Lessons Learned **:** Although capable of comprehensively characterizing memory effects and high-order nonlinearities, the Volterra series model suffers from exponentially growing parameters with increasing orders, resulting in prohibitive implementation complexity.

- The Hammerstein model:

If only the diagonal terms are retained in the Volterra series and the influence of the orthogonal terms is ignored, the series can be expressed as a Hammerstein model. The Hammerstein model is a form of filter cascade that combines a filter representing memory effects with a power series model representing nonlinearity. The Structure of the Hammerstein model is shown in Fig. 3.3. This model simplifies the expression of the Volterra series, its mathematical model is as follows.

$$y(n) = \sum_{d=1}^{D-1} \sum_{l=0}^{L-1} h_{d,l} x(n-l) |x(n-l)|^{2d-2} \tag{3.3}$$

This configuration effectively reduces the complexity of the full Volterra series by focusing on a single nonlinear function following a linear memory system, which makes the Hammerstein model particularly useful for applications where the nonlinearity is predominantly concentrated after a linear dynamic effect.

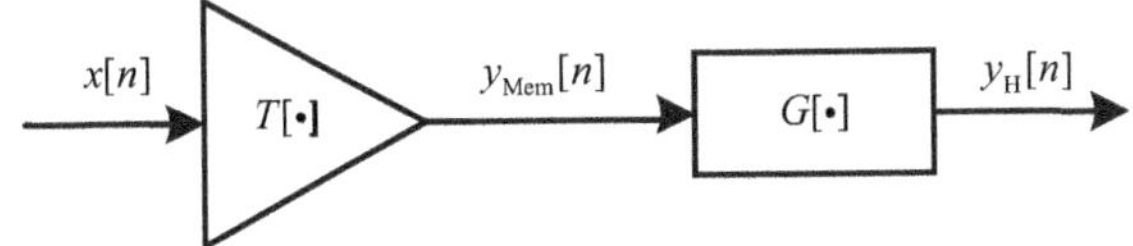

Fig. 3.3 Structure of the Hammerstein model

Reference [6] details an iterative method for identifying nonlinear systems from noisy input and output samples, tailored specifically for the Hammerstein model. This model incorporates a memoryless polynomial and a linear system, and iteratively adjusts the parameters of the linear system's transfer function and the polynomial's nonlinear coefficients to minimize mean square error. This approach has been effectively applied to estimate the channel responses of all nonlinear basis functions [7], successfully addressing nonlinear issues caused by IQ imbalance and power amplifier (PA) distortion. According to experimental results published in [8], this method achieved up to 46 dB of digital domain cancellation performance at a central frequency of 2.46 GHz and a bandwidth of 20 MHz. Furthermore, Ref. [9] proposed a Parallel Hammerstein (PH) structure for processing signals, which demonstrated superior cancellation effects under equivalent transmission power conditions. Additionally, the PH structure can independently address IQ imbalance and crosstalk issues in Multiple-Input Multiple-Output (MIMO) systems [10]. Despite the Hammerstein model's effectiveness in nonlinear self-interference cancellation, the multitude of highly correlated basis functions can potentially diminish the cancellation performance.

Lessons Learned **:** By cascading a linear filter with a power series, the Hammerstein model reduces complexity, yet its highly correlated basis functions may degrade interference cancellation performance.

- The Wiener model:

The Wiener model, as shown in Fig. 3.4, is a commonly used model for describing memory nonlinear systems. It consists of a linear filter and a power series model arranged in series, and can be considered a simplified form of the Volterra model. The linear filter component captures the memory effects of the system, while the power series model is employed to model the memoryless nonlinearities. Together, these two components form the Wiener model, effectively integrating the linear dynamic behavior with nonlinear transformations.

The mathematical expression of the Wiener model is as follows:

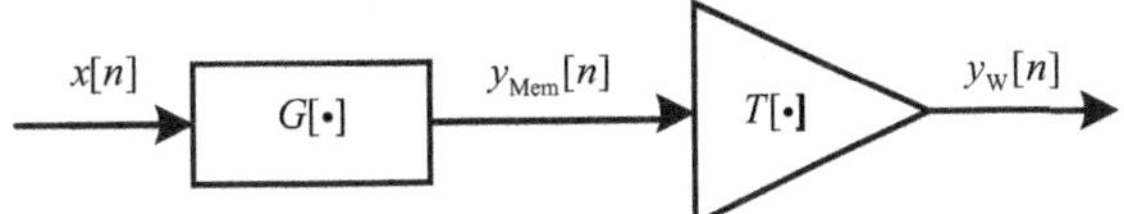

Fig. 3.4 Structure of the Wiener model

$$
\begin{aligned}
y(n) &= \sum_{k=0}^{K} a_k \left(\sum_{l=0}^{L-1} h_l x(n-l) \right)^{2k+1} \\
&= \sum_{k=0}^{K} a_k \left[\sum_{l=0}^{L-1} h_l x(n-l) \right] \left| \sum_{l=0}^{L-1} h_l x(n-l) \right|^{2k}
\end{aligned} \tag{3.4}
$$

Lessons Learned **:** The Wiener model decoupled linear-nonlinear structure enhances practicality, but its reliance on adaptive iterative algorithms constrains operational flexibility.

The Wiener and Hammerstein models are favored for their low complexity and practicality in implementation, showing improved performance compared to traditional nonlinear models. However, due to their simplistic structures, achieving highly accurate fits for complex nonlinear behaviors can be challenging. Some scholars have proposed solutions such as hybrid structures combining Wiener and Hammerstein elements. This approach is advantageous because it can capture the memory effects of the transmission channel before the power amplifier (PA), thereby not only mitigating PA-induced nonlinear distortions but also eliminating memory interference from the channel itself. Research in [11] demonstrated significant improvements in signal-to-noise ratio on in-band full-duplex (IBFD) relay nodes with various PA back-off levels through simulations. Moreover, this model is also frequently employed for frequency domain identification of nonlinear models; reference [12] implemented spline interpolation within this framework, proposing an effective digital predistortion method to compensate for high-power amplifiers, significantly reducing self-interference components and enhancing performance in self-interference cancellation. Reference [13] explored frequency domain identification algorithms for the Wiener-Hammerstein nonlinear model, employing polynomial functions or spline functions for approximating nonlinearity. Reference [12] developed both Wiener and Wiener-Hammerstein models using uniform spline interpolation as the basic nonlinearity function, applying FIR filtering to the linear blocks, and derived a gradient descent-based adaptive algorithm. Compared to polynomial models, these models reduce computational complexity by 68% and 53%, respectively. The Wiener model, however, requires adaptive iterative algorithms for nonlinear self-interference cancellation, which may limit its applicability.

Beyond these models, the Memory Polynomial (MP) and Generalized Memory Polynomial (GMP) models should also be considered. The Memory Polynomial model, a simplification derived from the Volterra model, primarily consists of tap delay lines and polynomial functions. Reference [14] built a predistorter PA model considering the memory effects of power amplifiers and used the Least Squares (LS) method for nonlinear self-interference cancellation, validating the robustness of this method. Building on the MP model, the addition of cross-memory terms allows for the development of the Generalized Memory Polynomial model. Reference [15] utilized the GMP model for digital baseband predistortion to compensate for nonlinear effects, showing superior results compared to traditional nonlinear models.

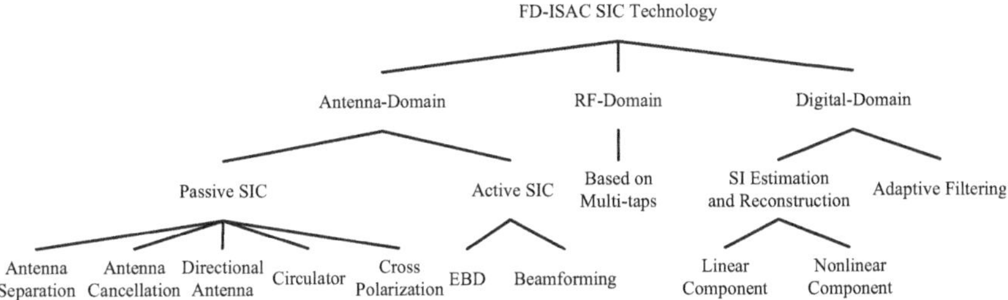

Fig. 3.5 FD SIC techniques

However, the more complex structure of the GMP model may increase the difficulty of processing in practical applications.

3.3 FD SIC Technologies

The architecture of FD-ISAC with three-domain cascade SIC is illustrated in Fig. 3.1. In practical FD-ISAC, combining with ASIC, RFSIC and DSIC technologies, the SI can be sufficiently eliminated to near the noise floor [16]. To be more specific, ASIC is achieved by isolating the transmit signal from the receive signal, while RFSIC and DSIC is achieved by accurately rebuilding a copy of SI and subtracting it from the received signal before and after the analog-to-digital converter (ADC), respectively. Importantly, compared with traditional FD communication system, the maximum delay of the TAPs in both RFSIC and DSIC needs to be limited to retain the components reflected from the environment or target [17]. In the following, the aforementioned SIC techniques will be investigated, as depicted in Fig. 3.5.

3.3.1 Antenna Domain SIC

(a) Passive ASIC: Passive ASIC leverages the inherent characteristics of antennas by designing their structure and layout to minimize the mutual influence between antennas, thereby increasing the isolation between them. Existing ASIC schemes are shown in Fig. 3.6 and summarized as follow.

- *Antenna Separation:* The path loss is increased by increasing the distance between the transceiver antennas or introducing isolation devices between them [18–20]. This method is simple to implement, but it has certain limitations, particularly in compact devices where achieving high levels of antenna isolation can be challenging due to space constraints.
- *Antenna Cancellation:* This method employs two transmit antennas and one receive antenna, strategically positioned to create a null in the interference pattern

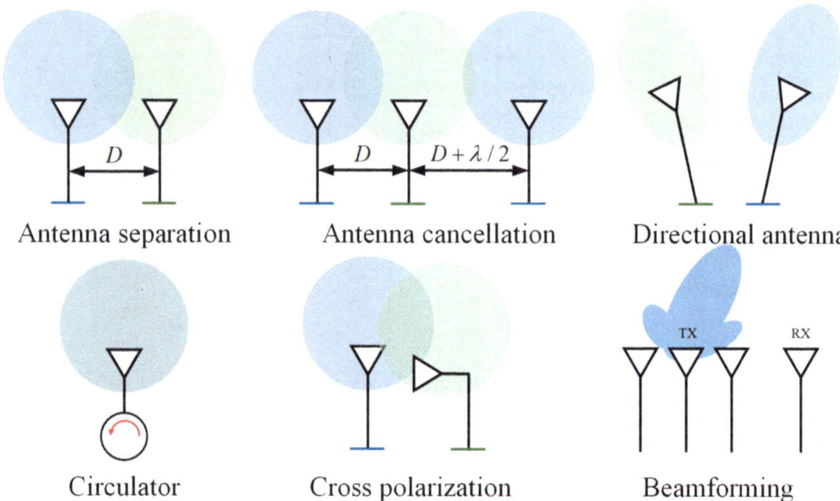

Fig. 3.6 Existing ASIC schemes

at the location of receive antenna. The transmitting antenna is placed in the distances D and $D + \lambda/2$ from the receiving antenna, where λ is the wavelength [21]. By adjusting the antenna distances, a phase difference of 180° is induced between the two transmitted signals at the receive antenna, resulting in destructive SIC.

- *Directional Antennas:* The directional approach minimizes the overlap of the radiation beams in space. The isolation between antennas can be further enhanced by applying a specified fixed beam spacing, such as transmitting and receiving signals in opposite directions [22].
- *Circulator:* The non-reciprocal characteristic of circulators can provide a certain level of isolation for the transmit and receive links. A circulator typically has three ports that connect the transmit link, the antenna and the receive link. The transmit link signal enters the antenna through the circulator and is transmitted, and signals received by the antenna is coupled into the receive link [23].
- *Cross-Polarization:* The transmitting and receiving antennas use orthogonal polarization to achieve SI suppression in the spatial domain. When the polarization of the transmit and receive antennas is orthogonal, theoretically the receive antenna will not receive signals from the transmit antenna, resulting in infinite isolation [24]. The simplest way to implement dual-polarized antennas is to place two identical radiating elements perpendicularly to each other. The symmetry of the two polarizations allows for lower port coupling, thereby enhancing SI suppression capability [25, 26].

(b) Active ASIC: Different from passive ASIC, active ASIC technology rely on tuning impedance or matching components in the antenna coupling network or controlling equipment. In addition, when there are multiple transmitting antennas, transmitters, and associated degrees of freedom, the beam direction of the antennas can be adjusted

to flexibly tailor the SIC to the actual situation. These active ASIC techniques are described in more detail within this section.

- *Electrical-balance Duplexer (EBD):* An EBD consists of a hybrid transformer and a balanced network, forming a lossless, reciprocal four-port system designed to isolate opposing ports. Assuming all ports are connected with matched loads, an input signal at any port will be transmitted to the adjacent two ports without coupling to the opposite port. This isolation property makes the EBD suitable for isolating transmit and receive signals in FD-ISAC systems with the advantages of compact size, high linearity and low insertion loss [27, 28]. Minor variations in antenna impedance can significantly impact the SIC performance. To address changes in antenna impedance, optimization algorithms such as adaptive filtering should be explored to enable the EBD to optimally match the network impedance. In 2019, [29] optimized the EBD matching network impedance using the Particle Swarm Optimizer (PSO) algorithm and applied the optimization scheme to the FD-ISAC architecture. However, the system operated at Sub-6 GHz frequencies with limited bandwidth and focused solely on speed measurement, neglecting distance measurement and higher frequency bands such as millimeter waves. In 2021, [30] implemented a FD-ISAC transceiver architecture using a USRP board with FPGA, achieving real-time optimization of the EBD matching network impedance based on DLS algorithm within 0.125 ms. However, similar to [29, 30] also faced limitations in terms of frequency range, bandwidth, and distance measurement capability.
- *Beamforming:* Joint beamforming at the transmitter and receiver aligns the null of the transmit antenna with the main lobe of the receive antenna, realizing the dual function of communication and sensing by reasonably allocating beam resources. This approach leverages the directivity of the beam to separate the SI from the desired signal, thereby mitigating interference and enhancing system performance. For MIMO systems, [31] proposed a novel framework for the joint optimization design of A/D transmit and receive beamformers as well as SIC. The goal is to find the digital beamforming matrix and the SIC matrix to maximize the achievable downlink rate and the accuracy of radar target sensing performance, while suppressing the SI signal power below the RF saturation level. Specifically, for the mmWave channel model, the proposed FD-ISAC system is able to estimate the relative velocities of all six targets with an estimation error of less than 1.5%. Reference [17] also discusses the joint design of transmit and receive beamformers at the transceiver to optimize the following objectives. Firstly, maximize the uplink and downlink communication rates. After that, maximize the transmit and receive radar beam pattern power at the target. Finally, suppress residual SI. An iterative algorithm based on penalty is proposed, which can effectively achieve up to 60 dB of SI suppression.

  Lessons Learned **:** Single-antenna ASIC architectures enable compact systems with simplified implementation. However, limited isolation (< 30 dB) between transmit/receive paths persists in both EBD and circulators due to impedance mismatches, making optimized matching a critical challenge.

3.3.2 RF Domain SIC

RFSIC is frequently employed for suppressing SI before the receiver's input, so as to prevent saturation of ADCs. In particular, multi-taps RF canceller design has been proved to be the most effective way to suppress multipath SI, where each TAP include a fixed delay line, a attenuator and a PS. Moreover, vector modulators (VMs) can substitute attenuators and PSs to adjust both the arbitrary amplitude and phase, which can effectively reduce hardware complexity. In [23], a 3-TAP VM-based RF canceller was proposed for FD-ISAC device. A gradient-based learning algorithm was also designed to tune the weights of VMs, which can achieve more than 50 dB RF cancellation. Importantly, compared with existing RFSIC canceller, [23] only eliminates SI within 1.5 m, while retaining reflected echoes beyond 1.5 m. Similar to [29], the frequency band considered in [23] is also Sub-6 GHz, with a relatively small bandwidth.

***Lessons Learned*:** RF cancellers face a hardware-complexity-versus-cancellation-performance trade-off. Multi-Tap designs demand more taps for enhanced cancellation, yet incur increased circuit complexity and scale. Thus, cost-effective RF cancellers with high cancellation capability require further research.

3.3.3 Digital Domain SIC

Due to the limitations in device performance and engineering techniques, ASIC and RFSIC often struggle to reduce the SI to the noise floor level. Therefore, it is necessary to further eliminate the residual SI signals after the ADC through DSIC to further reduce the SI power [32]. The core idea of DSIC is similar to that of the RFSIC, which involves reconstructing a cancellation signal to suppress SI. Typically, the baseband transmit signal of the transmitter is used as a reference source. Digital signal processing techniques are employed to estimate the parameters of the SI channel and then reconstruct the SI signal. Common DSIC techniques in FD-ISAC systems include two types: SI estimation and reconstruction, and adaptive filtering.

(a) SI Estimation and Reconstruction: The DSIC method based on SI estimation and reconstruction utilizes the baseband digital signal from the local transmitter and the received SI signal to estimate the SI channel. After reconstructing the SI in the digital-domain, it is subtracted from the received signal to suppress the SI.

In 2019, [29] implemented a DSIC module on a field-programmable gate array (FPGA) that models the multipath channel and hardware non-ideal factors, estimating the intended received signal and its reflection. These estimated signals are subtracted from the baseband received signal to obtain the remaining perceptual SI signal. In 2020, [30] enhanced the modeling of SI by incorporating a Hammerstein model for static nonlinearity and utilized an adaptive linear filter to track the linear components of the SI signals. The Least Mean Square (LMS) optimization algorithm was employed to dynamically update the parameters of both the Hammerstein model and the adaptive filter. In 2024, [33] extended the research to the mmWave (30 GHz), deriving a more accurate tapped channel model to characterize nonlinear SI by considering fractional delays. A nonlinear SIC suitable for beam tracking was proposed, achieving approximately 28 dB cancellation.

(b) Adaptive Filtering: This method utilizes the baseband transmit signal to obtain a SI reference signal through adaptive filtering. It calculates the error between the SI reference signal and the actual received signal, iteratively updating the filter coefficients using adaptive algorithms such as LMS and Recursive Least Squares (RLS) to minimize the error signal, ensuring effective SIC.

In 2019, [23] employed a novel self-orthogonal learning rule to dynamically estimate the parameters of the adaptive filter, thus avoiding explicit basis function orthogonalization and significantly reducing the computational complexity of the digital-domain parameter estimation algorithm. The results indicated that after employing three-domain cancellation, the dynamic range of perception could be increased enabling the detection of static and dynamic targets at a distance of 40 m, with a distance estimation accuracy of 1m and a target detection probability exceeding 90%. In 2020, [35] proposed a design method for DSIC based on adaptive filters and implemented it with Matlab software with poor real-time performance. In 2021, [34] designed a 28-taps adaptive filter in the digital-domain, utilizing LMS algorithm to update the tap coefficients in real-time. The DSIC algorithm, which was implemented in Matlab software as [35] was ported to an FPGA hardware platform, leading to the design and realization of an ISAC prototype. With a bandwidth of 40 MHz, the system was capable of achieving approximately 40 dB of RFSIC and 30 dB of DSIC. Additionally, building upon velocity measurement capabilities, the system incorporated distance measurement functionality, enabling the detection of moving targets within the range of 5–25 m.

(c) Intelligent methods: Intelligent methods primarily leverage artificial intelligence technologies to train and eliminate digital domain nonlinear self-interference components, with machine learning being a mainstream approach that has garnered widespread attention in academia. Deep learning (DL), a significant branch of machine learning, consists of multiple layers that can process and abstract nonlinear information at various levels, extracting effective feature functions from data models. The foundational architecture of deep learning is the artificial neural network (ANN).

Deep learning network architectures can be categorized into three types: Deep Feedforward Networks (DFN), Convolutional Neural Networks (CNN), and Recurrent Neural Networks (RNN). These networks are capable of modeling and analyzing unknown systems based on training data, making them suitable for characterizing complex system models. This advantage also facilitates their application in complex communication systems. As traditional communication algorithms have matured with increasingly limited scope for enhancement, deep learning offers new avenues for innovation. However, the complexity of deep learning techniques still needs to be reduced to levels acceptable for practical communication systems. Reference [36] introduced a real-time nonlinear self-interference cancellation scheme, where the self-interference channel is modeled by a deep feedforward neural network. This study demonstrated the feasibility of applying neural networks for full-duplex digital domain self-interference suppression, achieving 17 dB of digital cancellation with an average bit error rate of 8.5%. Experiments in Ref. [37] showed that when implementing dedicated integrated circuits, the hardware efficiency was 81% higher than that of memory polynomial model suppression schemes.

Reference [38] proposed a nonlinear SI cancellation method using a DFN network to construct the nonlinear part of the digital cancellation signal, verified with hardware measurements. The results indicated that the cancellation performance of the neural network-based method was equivalent to that of traditional polynomial-based methods, but with 36% fewer multiplications required and no need to compute any nonlinear basis functions.

Reference [39] studied various neural networks and compared their computational requirements with traditional methods. The findings revealed that at the same level of self-interference cancellation, such as 44 dB, RNN required 30.5% fewer floating-point operations than the polynomial model.

Reference [40] presented the concept of deep unfolding to model cascaded RF nonlinear systems. By using backpropagation to adjust complex-value parameters, this approach models a full-duplex self-interference cancellation system with IQ imbalance and PA nonlinearity. By using appropriate nonlinear models for each nonlinearity and cascading them, the overall model is simplified. For about 44.5 dB of self-interference cancellation performance, the number of model parameters could be reduced by 74%, and the number of operations per sample could be decreased by 79% compared to the extended linear parameter polynomial model.

In conclusion, Intelligent methods are still in the exploratory stage, and although they are capable of model parameter simplification, the elimination results are heavily dependent on the data quality of the training sequences and cannot be adaptively eliminated according to channel variations, so they have not yet been applied on a large scale.

Lessons Learned **:** Existing closed-form estimators for SI channel parameters exhibit limited adaptability in dynamic environments, while iterative alternatives remain vulnerable to suboptimal convergence due to initialization dependencies. Future research should focus on robust estimation frameworks that jointly optimize dynamic tracking stability and convergence robustness, potentially leveraging cross-layer optimization between hardware reconfigurability and algorithmic adaptability for real-time FD-ISAC systems.

3.4 Conclusion

In this chapter, we introduce the Full-Duplex (FD) system model, including its architecture and self-interference (SI) models (covering both linear and nonlinear characteristics). It further elaborates on interference cancellation techniques across the antenna, radio frequency (RF), and digital domains, laying a technical foundation for subsequent FD integrated communication-sensing systems.

References

1. Korpi D, Riihonen T, Anttila L, Valkama M (2018) Self-interference modeling and digital cancellation along with full-duplex wireless system analysis. In: SPCOM - international conference on signal processing communication, pp 432–436
2. de Figueiredo RJP (1982) The Volterra and Wiener theories of nonlinear systems. Proc IEEE 70(3):316–317
3. Zhu A, Brazil TJ (2004) Behavioral modeling of RF power amplifiers based on pruned volterra series. IEEE Microw Wirel Compon Lett 14(12):563–565
4. Islam MA, Smida B (2019) A comprehensive self-interference model for single-antenna full-duplex communication systems. In: ICC 2019 - IEEE international conference on communication, pp 1–7
5. Korpi D, Turunen M, Anttila L, Valkama M (2018) Modeling and cancellation of self-interference in full-duplex radio transceivers: Volterra series-based approach. In: IEEE international conference on communication workshops, pp 1–6
6. Narendra K, Gallman P (1966) An iterative method for the identification of nonlinear systems using a Hammerstein model. IEEE Trans Autom Control 11(3):546–550
7. Heino M et al (2015) Recent advances in antenna design and interference cancellation algorithms for in-band full duplex relays. IEEE Commun Mag 53(5):91–101
8. Anttila L et al (2014) Modeling and efficient cancellation of nonlinear self-interference in MIMO full-duplex transceivers. In: IEEE globecom workshops, pp 777–783
9. Anttila L et al (2013) Cancellation of power amplifier induced nonlinear self-interference in full-duplex transceivers. In: Asilomar conference on signals systems and computers, pp 1193–1198
10. Komatsu K, Miyaji Y, Uehara H (2018) Basis function selection of frequency-domain Hammerstein self-interference canceller for in-band full-duplex wireless communications. IEEE Trans Wirel Commun 17(6):3768–3780
11. Gregorio FH et al (2017) Predistortion for power amplifier linearization in full-duplex transceivers without extra RF chain. In: IEEE international conference on acoustic speech signal processing, pp 6563–6567
12. Pascual Campo P et al (2021) Cascaded spline-based models for complex nonlinear systems: methods and applications. IEEE Trans Signal Process 69:370–384
13. Sano M, Sun L (2002) Identification of Hammerstein-Wiener system with application to compensation for nonlinear distortion. Proc 41st SICE Annu Conf 3:1521–1526
14. Ding L et al (2004) A robust digital baseband predistorter constructed using memory polynomials. IEEE Trans Commun 52(1):159–165
15. Morgan DR et al (2006) A generalized memory polynomial model for digital predistortion of RF power amplifiers. IEEE Trans Signal Process 54(10):3852–3860
16. Kolodziej KE, Perry BT, Herd JS (2019) In-band full-duplex technology: techniques and systems survey. IEEE Trans Microw Theory Tech 67(7):3025–3041
17. Liu Z et al (2023) Joint transmit and receive beamforming design in full-duplex integrated sensing and communications. IEEE J Sel Areas Commun 41(9):2907–2919
18. Duarte M, Dick C, Sabharwal A (2012) Experiment-driven characterization of full-duplex wireless systems. IEEE Trans Wirel Commun 11(12):4296–4307
19. Syrjala V et al (2014) Analysis of oscillator phase-noise effects on self-interference cancellation in full-duplex OFDM radio transceivers. IEEE Trans Wirel Commun 13(6):2977–2990
20. Duarte M, Sabharwal A (2010) Full-duplex wireless communications using off-the-shelf radios: Feasibility and first results. In: Asilomar conference on signals, systems, and computers, pp 1558–1562

21. Choi JI et al (2010) Achieving single channel, full duplex wireless communication. In: Proceedings of annual international conference on mobile computing and networking, pp 1–12
22. Everett E, Sahai A, Sabharwal A (2014) Passive self-interference suppression for full-duplex infrastructure nodes. IEEE Trans Wirel Commun 13(2):680–694
23. Baquero Barneto C et al (2019) Full-duplex OFDM radar with LTE and 5G NR waveforms: challenges, solutions, and measurements. IEEE Trans Microw Theory Tech 67(10):4042–4054
24. Debaillie B et al (2014) Analog/RF solutions enabling compact full-duplex radios. IEEE J Sel Areas Commun 32(9):1662–1673
25. Zhang Z et al (2016) Full-duplex wireless communications: challenges, solutions, and future research directions. Proc IEEE 104(7):1369–1409
26. Dai L et al (2015) Non-orthogonal multiple access for 5G: solutions, challenges, opportunities, and future research trends. IEEE Commun Mag 53(9):74–81
27. van Liempd B et al (2016) A +70-dBm IIP3 electrical-balance duplexer for highly integrated tunable front-ends. IEEE Trans Microw Theory Tech 64(12):4274–4286
28. Laughlin L et al (2016) Passive and active electrical balance duplexers. IEEE Trans Circuits Syst II Exp Briefs 63(1):94–98
29. Hassani SA et al (2019) Doppler radar with in-band full duplex radios. In: Proceedings of IEEE INFOCOM, pp 1945–1953
30. Hassani SA et al (2021) In-band full-duplex radar-communication system. IEEE Syst J 15(1):1086–1097
31. Islam MA, Alexandropoulos GC, Smida B (2022) Integrated sensing and communication with millimeter wave full duplex hybrid beamforming. In: IEEE international conference on communication, pp 4673–4678
32. Ahmed E, Eltawil AM (2015) All-digital self-interference cancellation technique for full-duplex systems. IEEE Trans Wirel Commun 14(7):3519–3532
33. He Y, Zhao H, Shao S (2024) Nonlinear self-interference cancellation in vehicle networks for full-duplex integrated sensing and communication. IEEE Trans Veh Technol 73(9):13980–13985
34. Hassani SA et al (2022) Joint in-band full-duplex communication and radar processing. IEEE Syst J 16(2):3391–3399
35. Hassani SA et al (2021) Adaptive filter design for simultaneous in-band full-duplex communication and radar. In: European radar conference, pp 5–8
36. Guo H et al (2019) DSIC: deep learning based self-interference cancellation for in-band full duplex wireless. In: IEEE global communication conference, pp 1–6
37. Kurzo Y, Burg A, Balatsoukas-Stimming A (2018) Design and implementation of a neural network aided self-interference cancellation scheme for full-duplex radios. In: Asilomar conference on signals, systems, and computers, pp 589–593
38. Balatsoukas-Stimming A (2018) Non-linear digital self-interference cancellation for in-band full-duplex radios using neural networks. In: IEEE SPAWC, pp 1–5
39. Kristensen AT, Burg A, Balatsoukas-Stimming A (2019) Advanced machine learning techniques for self-interference cancellation in full-duplex radios. In: Asilomar conference on signals, systems, and computers, pp 1149–1153
40. Kristensen AT, Burg A, Balatsoukas-Stimming A (2020) Identification of non-linear RF systems using backpropagation. In: IEEE ICC workshops, pp 1–6

cc
BY NC ND

Part II
Algorithm and Implementation of FD-ISAC

Chapter 4
FD-ISAC Processing

Abstract Compared to the traditional TDD mode, the FD mode is more suitable for ISAC systems since it allows simultaneous transmission and reception of signals (including both echoes and communication signals). Under the FD-ISAC framework, a completely new receiver signal flow needs to be designed to handle complex receiver signals. In FD-IBFD, the processing for communication signals is the same as the FD chapter process, and the radar echo signal is regarded as the self-interference signal for elimination. While the separation of the echo signal from the self-interference signal is the key to realize the accurate reception and processing of the sensed signal. And then the extracted echo signals are then subjected to signal processing to obtain sensing information.

4.1 FD-ISAC System Model

The conventional single-base station ISAC scheme adopts time-division duplex (TDD) mode to implement the transmission of ISAC signals and the reception of echoes. Compared with ISAC in full-duplex (FD) mode, as shown in Fig. 4.1, TDD mode has the following shortcomings in communication and sensing performance respectively.

- Communication performance: The TDD-based ISAC framework needs to reserve time slots for echo signal reception after downlink signal transmission, which inevitably increases the channel occupancy time. This time allocation loss reduces the system throughput and introduces additional communication delay.
- Sensing performance: Compared to radar pulses, a communication-centric ISAC system transmits downlink ISAC signals continuously over a relatively long period of time. With TDD mode, it will not be able to receive the echo signals from nearby targets during transmission, thus creating an obvious sensing blind zone in short-range sensing scenarios.

Therefore, the integrated signal transmission and echo reception based on the FD mode will simultaneously improve the communication and sensing performance of ISAC. However, FD-ISAC exists a great challenge which is the extraction of echo

C. Du et al., *Full-Duplex Integrated Sensing and Communication Systems*,
https://doi.org/10.1007/978-981-92-0470-0_4

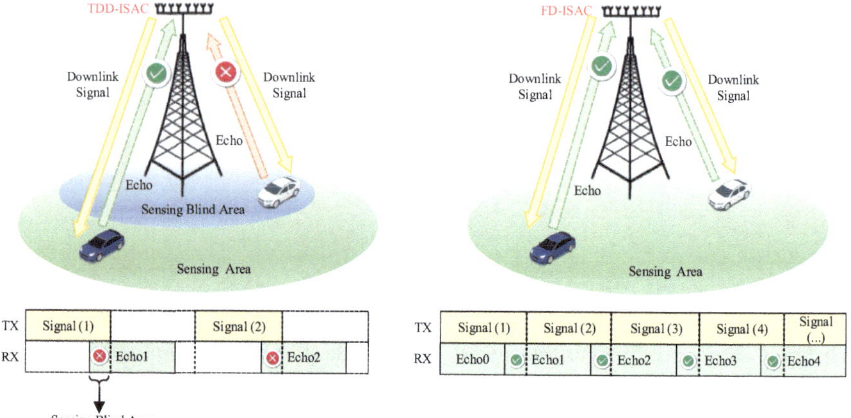

Fig. 4.1 TDD verses FD ISAC signal transmitting and receiving diagram

signals and the elimination of SI signals. This SI comprises two dominant components: (a) direct leakage interference, accounting for the largest power contribution due to insufficient antenna isolation between adjacent transceivers, and (b) multipath-reflected interference caused by the echo interference reflected from scatterers in the environment. The receiver consequently observes a mixture signal containing both target echoes and SI components, necessitating a fundamental redesign of signal processing procedure compared to conventional TDD-ISAC systems.

We consider a FD-ISAC system, the downlink transmitted signal is an ISAC signal, which is used to convey information to downlink users and to perform sensing tasks for target detection. In practical scenarios, due to the close proximity of the transmit and receive antennas, the transmitted signal is also received by the receiving antenna at the same node, resulting in SI signals. This means that the FD BS receives a mixed signal that contains the SI and echo signals from the target. The signal received at the FD BS can be expressed as

$$\mathbf{y}[n] = \mathbf{y}_s[n] + \mathbf{y}_{si}[n] + \mathbf{z}_r[n] = \underbrace{\mathbf{h}_s\mathbf{x}[n]}_{\text{Echo signal}} + \underbrace{\mathbf{h}_{si}\mathbf{x}[n]}_{\text{SI}} + \mathbf{z}_r[n] \tag{4.1}$$

where $\mathbf{y}_s[n] = \mathbf{h}_s\mathbf{x}[n]$ is the echo signal reflected by the target, $\mathbf{y}_{si}[n] = \mathbf{h}_{si}\mathbf{x}[n]$ refers to the SI signal, and $\mathbf{z}_r[n]$ is the additive white Gaussian noise.

As mentioned in the previous section, the high-power SI signals due to the short transmission distance, which will cause the analog-to-digital converter (ADC) saturation at the receiver and fail to detect the echo signals correctly. Therefore, the design of Self-Interference Cancellation (SIC) is especially critical for FD-ISAC. However, for FD-ISAC systems with sensing performance, the hardware design and algorithm implementation cannot completely reuse the existing communication FD SIC method, but need to specifically design FD SIC hardware and software solutions for the characteristics of ISAC, which can retain or reconstruct the target echo signal. Therefore, a reasonable design for the integrated RF transceiver architecture as well as key technical challenges such as SIC and echo signal extraction within the integrated system, must be addressed [1].

4.2 FD-ISAC Signal Processing Flow

When sending communication data, the FD-ISAC system needs to simultaneously detect the echo signals of its own transmitted signals hitting the target, so as to realize the function of wireless sensing. The SI and echo signals are received simultaneously, as shown in Fig. 4.2.

In traditional FD communication systems, target echo signals inherently manifest as multipath self-interference components, so the traditional SIC algorithm is ineffective for ISAC applications. Consequently, a novel signal processing framework is required to separate and extract target echoes from multipath interference while preserving sensing functionality. Therefore, the received signal processing flow of FD-ISAC is divided into the following steps:

- Step 1. Spatial-Domain Interference Mitigation: Direct-path interference is attenuated via spatial isolation techniques, leveraging antenna configuration and beamforming strategies to reduce interference power at the receiver front-end.
- Step 2. Analog-Domain Interference Mitigation: Residual direct-path interference is further suppressed using analog cancellation circuits, ensuring the aggregate interference remains below the receiver ADC's saturation threshold.
- Step 3. Separation of Radar Echo and SI: The novel algorithm applicable to ISAC disentangle target echoes from residual direct interference and uncanceled multipath components, employing adaptive signal decomposition and interference estimation techniques.
- Step 4. Echo signal processing: process the extracted echo signal to realize the perception function

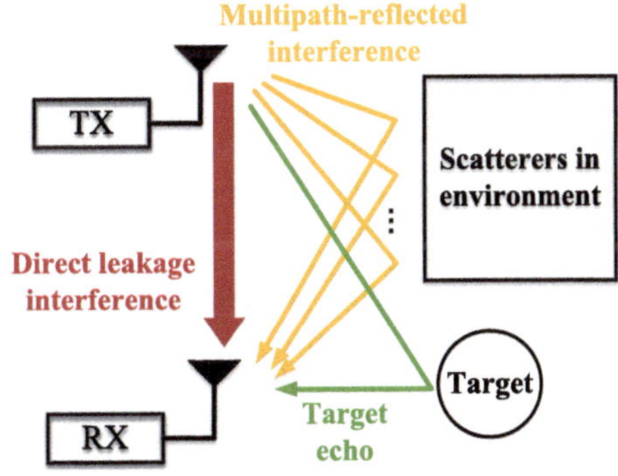

Fig. 4.2 FD-ISAC receiving signal diagram

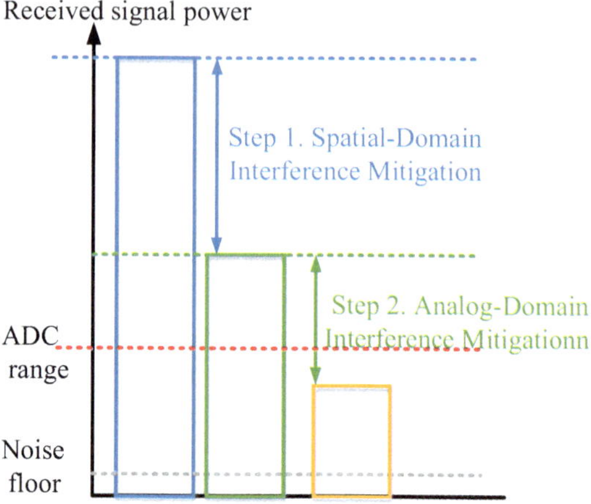

Fig. 4.3 FD-ISAC direct leakage interference cancellation schematic

The SIC mechanism in Steps 1 and 2 achieves self-interference suppression through spatial-RF domain cascade, reducing received signal power within the ADC quantization range as depicted in Fig. 4.3.

Residual self-interference and target echoes are subsequently processed by the receiver ADC. Since Chap. 3 comprehensively addresses SIC algorithms in FD communication systems, this chapter refrains from reiterating implementation details for these steps.

However, echo signal extraction in FD-ISAC systems presents unique challenges compared to conventional FD communication. In traditional FD systems, all received signals originating from the transmitter—including direct antenna leakage and environmental reflections are classified as SI. In contrast, FD-ISAC systems differentiate between two distinct components: (a) Self-Interference: Direct transmitter-to-receiver leakage and multipath reflections from static environmental objects. (b) Useful Signals: Target-reflected echoes carrying sensing information.

This critical distinction necessitates specialized signal separation strategies, as suppressing environmental multipath interference risks attenuating target echoes. The subsequent sections focus on overcoming this challenge through adaptive signal decomposition techniques that preserve sensing fidelity while maintaining SI suppression efficacy.

4.3 Separation of Target Echo and SI

After employing three-domain SIC, the echo will be extracted from the residual SI. In the following subsubsection, echo extraction method, including based on independent component analysis (ICA) [2], multi-tap analog domain cancellation [3], and Doppler-based dynamic target extraction [4], will be investigated.

(1) ICA-based Method: This paper proposes a two-stage signal recovery scheme for IBFD-ISAC systems. The first stage employs independent component analysis (ICA) for SIC, where the mixed signal is modeled as $Y = HS$ (H is the mixing matrix, S is the source signal). The demixing matrix W is used to separate the communication signal (SI) and the target radar reflection signal (RS), i.e., $\hat{Y} = WY$. The second stage detects the RS delay via cross-correlation operations, segmenting the signal into three parts to locate the peak in cross-correlation results for target distance extraction. Simulation results show a clear peak in the second-segment cross-correlation, verifying the effectiveness of delay detection. The output signal-to-interference-plus-noise ratio (OSINR) is comparable to that of conventional IBFD systems. When the delay exceeds a threshold, OSINR deteriorates significantly due to the disappearance of the cross-correlation peak, indicating sensitivity to delay ranges.

ICA's blind source separation effectively handles strong self-interference; the two-stage approach integrates signal separation and target parameter extraction. Cross-correlation improves delay detection accuracy for target distance calculation. Relies on known preamble sequences for data positioning, with unvalidated robustness to signal structure changes in complex multipath environments. The iterative nature of ICA may introduce computational latency, affecting real-time performance.

(2) Multi-tap analog domain cancellation: During the SIC process, by setting the maximum time delay of the taps in RF-domain and the tap length of the digital-domain adaptive filter, the echo corresponding to this time delay that is outside a certain distance can be retained. [3] designed an RF-domain multi-tap canceller, representing the signal after RF SIC as $y_{\mathrm{RF}}(t) = r_{\mathrm{RF}}(t) - \hat{s}_{\mathrm{RF}}(t)$, where $r_{\mathrm{RF}}(t)$ denotes the input signal to the receiver and $\hat{s}_{\mathrm{RF}}(t)$ denotes reconstructed signal. $\hat{s}_{\mathrm{RF}}(t)$ can be further obtained as

$$\hat{s}_{\mathrm{RF}}(t) = \sum_{l=1}^{N} \alpha_l \left(\cos(\theta_l) x_{\mathrm{RF}}^{0}(t - \tau_l) - \sin(\theta_l) x_{\mathrm{RF}}^{90}(t - \tau_l)\right) \tag{4.2}$$

where $x_{RF}(t)$ represents the output signal of the power amplifier, N denotes the number of taps, τ_l signifies the RF delay of the l(tap, and α_l indicate the adjustable amplitude and phase values of the l-th vector modulator, respectively. $x_{\mathrm{RF}}^{0}(t)$ and $\chi_{\mathrm{RF}}^{90}(t)$ correspond to the in-phase and quadrature components of the vector modulator. Taking the three-tap example, in the design of the RF canceller, the maximum delay is set to 10ns, which corresponds to an equivalent distance of 3m, preventing the cancellation of any echoes from real targets beyond a distance of 1.5m. Similarly, as for DSIC, the number of taps in the filter is directly related to the detectable sensing distance. In the cited literature, the DSIC canceller operates at a sampling rate of 240MHz, with the filter tap length set to 11, corresponding to a minimum detectable distance of approximately 3m, which matches the parameters of the RF canceller.

After separating the echoes from the SI signals in the received signals, the subsequent steps of SIC, communication signal demodulation, and finally, sensing estimation can be carried out.

However, due to the limitation of the number of taps in the analog domain, this paper can only realize the extraction of a limited number of different path signals within a certain distance, and cannot distinguish whether it is the multipath signal of the environment or the echo signal of the target.

(3) Doppler-based dynamic target extraction: The core principle of the proposed Frequency-Domain Differential Interference Cancellation (F-DIC) scheme is to leverage the time-invariance of static interference (direct self-interference and multipath reflections) across consecutive OFDM symbols for interference cancellation and moving target echo extraction. Static interference, generated by stationary objects, exhibits invariant channel responses between adjacent OFDM symbols, i.e., the static channel matrix satisfies $H^S_{n,m+1} = H^S_{n,m}$. For the received signal at the n-th subcarrier and m-th symbol,

$$R_{n,m} = H^M_{n,m} X_{n,m} + H^S_{n,m} X_{n,m} + N_{n,m}, \tag{4.3}$$

where $H^M_{n,m}$ is the moving target channel matrix, $H^S_{n,m}$ is the static interference channel matrix, and $N_{n,m}$ is the noise.

The differential operation

$$T_{n,m} = R_{n,m} - \frac{X_{n,m}}{X_{n,m+1}} R_{n,m+1}, \tag{4.4}$$

cancels the static interference term $H^S_{n,m} X_{n,m}$ due to its time-invariance, while retaining moving target signals whose channels $H^M_{n,m}$ change with symbols due to Doppler shifts. Through recursive smoothing and mean estimation, the output signal $\overline{R}_{n,m} = \hat{E}_{n,m}$ is obtained with static interference eliminated, achieving CSI-free interference cancellation. By utilizing the frequency-domain processing of OFDM systems and differential accumulation over N_s symbols, the method reduces computational complexity while enhancing sensing performance effectively.

The proposed F-DIC scheme can detect targets with velocities in the range $0m/s < speed \leq 75m/s$, where:

The maximum detectable velocity is 75 m/s, determined by the OFDM subcarrier orthogonality constraint $f_D \leq 0.2\Delta f$ (subcarrier spacing $\Delta f = 60$ kHz), calculated via $v_{\max} = \frac{c \cdot f_{D,\max}}{2 f_c}$ ($f_{D,\max} = 12$ kHz, carrier frequency $f_c = 24$ GHz).

The minimum detectable velocity is just above 0 m/s, the lower bound is theoretically limited by velocity resolution (2.68 m/s), but practically depends on target power and noise. Static targets (v = 0 m/s) are fully suppressed by F-DIC and undetectable.

The velocity resolution is 2.68 m/s, determined by the number of OFDM symbols ($N_s = 128$) using the formula $\Delta v = \frac{c}{2 N_s T_s}$, reflecting the system's ability to distinguish targets with different velocities.

The above three methods are existing techniques that can distinguish between self-interfering signals and echo signals in FD-ISAC. The next section will present in the aspect of sensing processing algorithms from the target echo signals extracted in this section.

4.4 Radar Signal Processing Algorithms

In FD-ISAC systems, sensing estimation algorithms are employed to estimate parameters such as distance and velocity after SIC. In the existed researches, there are mainly four categories for sensing estimation, i.e., two dimensional (2D)-DFT Periodogram, Subspace Spectral Analysis, Grid-Based Compressed Sensing (CS) and Neural Networks, as shown in Table 4.1.

4.4.1 2D-DFT Periodogram

The classical 2D-DFT is a widely used periodogram method in radar systems. It provides a rough estimation of sensing parameters by combining two of the following transformations: converting time-domain samples to the frequency domain, converting spatial samples to the angle domain, and converting phase samples to the

Table 4.1 Comparison of sensing estimation algorithms in ISAC systems

Algorithm	Estimated parameters	Characteristics	Limitations
Periodogram techniques(e.g., 2D DFT) [4–6]	Distance, Velocity	Traditional approach, easy to implement. Can be a starting point for other algorithms.	Low resolution. Generally requires a complete set of continuous samples in all domains, which may not always be available
Subspace method(e.g., MUSIC) [7]	Distance, Velocity, DOA	High resolution, can estimate off—grid parameters.	High complexity
Compressive Sensing on Grid [9–11]	Distance, Velocity	Flexible, does not require continuous sampling. Allows for various recovery algorithms to balance complexity and performance.	Significant performance degradation when there are many continuous parameter paths
Deep Learning [13–15]	Distance, Velocity	High resolution, robust for multi - target estimation.	Requires a large amount of training data, and training is complex

Doppler frequency domain. Pure channel information is extracted by computing the ratio between transmitted and received symbols, thereby eliminating communication data:

$$s(m,n) = \frac{x(m,n)}{y(m,n)} = H(m,n) \cdot (\overline{k}_r \otimes \overline{k}_d)_{m,n} \tag{4.5}$$

where:

- $\overline{k}_r$: Range-induced phase shift vector (frequency-dependent):

$$\overline{k}_r = \left[e^{-j2\pi f_1 \frac{2R_r}{c}}, e^{-j2\pi f_2 \frac{2R_r}{c}}, \ldots, e^{-j2\pi f_N \frac{2R_r}{c}} \right]^T \tag{4.6}$$

- $\overline{k}_d$: Doppler-induced phase shift vector (time-dependent):

$$\overline{k}_d = \left[1, e^{j2\pi T_{\mathrm{OFDM}} f_{d,r}}, \ldots, e^{j2\pi (M-1) T_{\mathrm{OFDM}} f_{d,r}} \right] \tag{4.7}$$

- $\otimes$: Kronecker product jointly representing range and Doppler effects

The channel information matrix $s(m,n)$ undergoes dual Fourier transformations:

1. **Row-wise processing (DFT):** Applied per row (fixed m, varying n) to extract range information, where phase differences across subcarriers reflect target range.
2. **Column-wise processing (IDFT):** Applied per column (fixed n, varying m) to extract Doppler information, where inter-symbol phase variations indicate target velocity.

In [4], an OFDM-based ISAC system was proposed, where DFT is applied to the received signal to obtain Doppler estimates and the IDFT is used for delay estimation. However, since the estimation of distances and speeds are discrete, both of them are subject to quantization errors. To solve this issue, in [5], rough expressions for ranging and velocity accuracy were derived and two algorithms, Fractional Fourier Transform (FRFT) and phase analysis, were proposed to reduce quantization noise. Moreover, in [6], precision expressions for velocity estimation were proposed to further improve the estimation accuracy.

However, the inherent since function tail in DFT leads to lower resolution and continuous measurements in either the time or frequency domain were required, which limits the application of 2D-DFT Periodogram.

4.4.2 *Subspace Spectral Analysis*

Classic subspace-based spectral analysis techniques, such as multiple signal classification (MUSIC) and estimating signal parameter via rotational invariance techniques (ESPRIT), allow for high-resolution estimation of continuous-valued parameters [7].

Both methods rely on subspace decomposition for localization, with MUSIC utilizing the noise subspace and ESPRIT leveraging the signal subspace.

(1) MUSIC Algorithm: The fundamental principle of the MUSIC algorithm involves performing eigen-decomposition on the covariance matrix of received signal data to obtain the signal and noise subspaces. After that, spatial spectral function was constructed based on the parameters of the target source to estimate the Direction of Arrival (DOA). In [8], a 2D MUSIC algorithm was proposed to yield higher resolution estimates in OFDM-ISAC systems.

For MUSIC-based parameter estimation in OFDM-ISAC systems, consider K far-field narrowband signals incident on an M-element Uniform Linear Array (ULA). The received signal is modeled as:

$$\mathbf{x}(t) = \mathbf{A}(\theta)\mathbf{s}(t) + \mathbf{n}(t), \tag{4.8}$$

where $\mathbf{A}(\theta) = [\mathbf{a}(\theta_1), \ldots, \mathbf{a}(\theta_K)] \in \mathbb{C}^{M\times K}$ denotes the array manifold matrix with steering vector $\mathbf{a}(\theta_k) = [1, e^{j2\pi d \sin\theta_k/\lambda}, \ldots, e^{j2\pi(M-1)d\sin\theta_k/\lambda}]^T$ (element spacing d, wavelength λ, arrival angle θ_k), $\mathbf{s}(t) \in \mathbb{C}^{K\times 1}$ represents the source signal vector, and $\mathbf{n}(t) \in \mathbb{C}^{M\times 1}$ is additive white Gaussian noise.

The sample covariance matrix

$$\hat{\mathbf{R}} = \frac{1}{T}\sum_{t=1}^{T} \mathbf{x}(t)\mathbf{x}^H(t) = \mathbf{A}(\theta)\mathbf{R}_s\mathbf{A}^H(\theta) + \sigma^2\mathbf{I}_M, \tag{4.9}$$

where $\mathbf{R}_s = E[\mathbf{s}\mathbf{s}^H]$ and σ^2 is the noise power.

Perform eigendecomposition on the covariance matrix $\mathbf{R}$:

$$\hat{\mathbf{R}} = \mathbf{U}_s\mathbf{\Lambda}_s\mathbf{U}_s^H + \mathbf{U}_n\mathbf{\Lambda}_n\mathbf{U}_n^H. \tag{4.10}$$

This partitions the space into: signal subspace $\mathbf{U}_s \in \mathbb{C}^{M\times K}$ (spanned by large-eigenvalue eigenvectors) and noise subspace $\mathbf{U}_n \in \mathbb{C}^{M\times(M-K)}$ (spanned by small-eigenvalue eigenvectors), satisfying $\mathbf{U}_s^H\mathbf{U}_n = \mathbf{0}$.

The MUSIC spectrum

$$P_{\text{MUSIC}}(\theta) = 1/\mathbf{a}^H(\theta)\mathbf{U}_n\mathbf{U}_n^H\mathbf{a}(\theta), \tag{4.11}$$

exhibits peaks at true AoAs θ_k where $\mathbf{a}(\theta_k) \perp \mathbf{U}_n$.

For Direction of Arrival (DoA) estimation, the spatial spectrum function is constructed using echo signals received by different antenna array elements. For range and velocity estimation, the channel information matrix is first formulated according to the method in Sec.4.4.1. Subsequently, the target's range is estimated by exploiting phase differences across subcarriers, while its velocity is determined through phase variations between OFDM symbols. Specifically, processing the row and column vectors of this channel information matrix enables the construction of spatial spectrum functions, whose peak locations are then searched to estimate the target's range and velocity parameters.

(2) ESPRIT Algorithm: ESPRIT algorithm divides a uniform array into two overlapping sub-arrays, estimating the DOA of incoming signals through the rotational invariance relationship between the two sub-arrays. In [7], an ESPRIT algorithm for automatic pairing was proposed to achieve super-resolution for distance and velocity estimation in in OFDM-ISAC systems.

The deterministic signal components (cisoids) are separated from stochastic noise by constructing time-shifted data matrices:

$$\mathbf{X} = \left[\mathbf{x}[0]\ \mathbf{x}[1] \cdots \mathbf{x}[K-2]\right], \quad \mathbf{Y} = \left[\mathbf{x}[1]\ \mathbf{x}[2] \cdots \mathbf{x}[K-1]\right] \tag{4.12}$$

where $\mathbf{X}$ contains the first $K-1$ samples and $\mathbf{Y}$ contains the last $K-1$ samples.

Noise variance σ^2 is estimated through the smallest eigenvalue of the autocorrelation matrix:

$$\mathbf{R}_{XX} = \frac{1}{K-1}\mathbf{X}\mathbf{X}^H, \quad \sigma^2 = \lambda_{\min}(\mathbf{R}_{XX}) \tag{4.13}$$

Modified covariance matrices are constructed to eliminate noise effects:

$$\mathbf{C}_{XX} = \mathbf{R}_{XX} - \sigma^2\mathbf{I} \tag{4.14}$$

$$\mathbf{C}_{XY} = \mathbf{R}_{XY} - \sigma^2\mathbf{Z} \tag{25}$$

where $\mathbf{I}$ is the identity matrix and $\mathbf{Z}$ is the shift operator matrix.

The estimation methods for DoA, range, and velocity follow an identical processing framework to the MUSIC approach described in Sect. 4.4.2.

Although subspace spectral analysis alleviates the limitations of spectral leakage in DFT, higher computational complexity is required. Moreover, the sensing accuracy of MUSIC depends on search granularity, and ESPRIT generally requires evenly spaced samples. To achieve high resolution, both methods often necessitate extensive sampling, which may not always be feasible in certain domains, such as spatial, requiring a large number of antennas.

4.4.3 Compressive Sensing

The CS technique transforms the parameter estimation problem into a sparse signal recovery problem, which can be solved by various algorithms such as convex relaxation, greedy algorithm and probabilistic inference.

Based on the on-grid CS technique the sensing parameters to be estimated include delays, AOA, and Doppler across three distinct domains. Given the relative independence of signals in these domains, they can be represented as high-dimensional (3D) vectors, or even in tensor form, enabling the application of 1D to 3D CS techniques in ISAC systems [9].

The three-dimensional received signal model is expressed as:

$$\begin{aligned} Y_{n,t} &= \sum_{\ell=1}^{L} b_\ell e^{-j2\pi n\tau_\ell f_0} e^{j2\pi t f_{D,\ell} T_s} \cdot \mathbf{a}(M_2, \phi_\ell)\mathbf{a}^T(M_1, \theta_\ell) + z_{n,t} \\ &= \mathbf{A}_{\mathrm{rx}}\mathbf{C}_n\mathbf{D}_t\mathbf{A}_{\mathrm{tx}}^T + z_{n,t}, \end{aligned} \tag{4.15}$$

where ϕ_ℓ denotes AoA, θ_ℓ represents AoD, and $\mathbf{a}(M,\theta) = [1, e^{j\pi\sin\theta}, \ldots, e^{j\pi(M-1)\sin\theta}]^H$ is the antenna array response vector for an M-element array. Here, $\mathbf{A}_{\mathrm{rx}}$ is the receive antenna matrix, $\mathbf{A}_{\mathrm{tx}}$ is the transmit antenna matrix, $\mathbf{C}_n$ is the delay diagonal matrix with elements $e^{-j2\pi n\tau_\ell f_0}$, $\mathbf{D}_t$ is the Doppler diagonal matrix with elements $b_\ell e^{j2\pi t f_{D,\ell} T_s}$, and $z_{n,t}$ is the additive noise matrix.

For 1D compressive sensing, stacking all subcarrier signals yields:

$$\mathbf{Y}_t = \mathbf{W} \cdot \mathbf{G}_t + \mathbf{Z}_t, \quad \mathbf{G}_t = \mathbf{D}_t\mathbf{A}_{\mathrm{rx}}^T, \tag{4.16}$$

where $\mathbf{W}$ is the delay dictionary matrix with the ℓ-th column $\{e^{-j2\pi n\tau_\ell f_0}\}$, and $\mathbf{G}_t$ combines Doppler and receive antenna information.

Treating $\mathbf{Y}_t$ as a grid-based Multiple Measurement Vector (MMV)(Multiple Measurement Vector (MMV)) problem with delay-domain sparsity:

$$\mathbf{Y}_t = \boldsymbol{\Psi}_1 \cdot \mathbf{X}_t + \mathbf{Z}_t, \tag{4.17}$$

where $\boldsymbol{\Psi}_1$ is a partial DFT delay dictionary approximating $\mathbf{W}$, and $\mathbf{X}_t$ is the sparse coefficient matrix. Solved via 1D Sparse Bayesian CS algorithms, non-zero element positions correspond to delays τ_ℓ.

AoA estimation uses column-wise cross-correlation of $\mathbf{G}_t = \mathbf{D}_t\mathbf{A}_{\mathrm{rx}}^T$:

$$\phi_\ell \approx \frac{1}{\pi}\angle\left(\sum_{p=1}^{M_2-1} (\mathbf{G}_t)^*_{:,p}(\mathbf{G}_t)_{:,p+1}\right), \tag{4.18}$$

where $(\mathbf{G}_t)_{:,p}$ denotes the pth column of $\mathbf{G}_t$.

Doppler estimation employs row-wise cross-correlation across multiple DMRS signals:

$$f_{D,\ell} \approx \frac{1}{2\pi T_s}\angle\left(\sum_{t=1}^{N_d-1} (\mathbf{G}_t)_{\ell,:}(\mathbf{G}_{t+1})^*_{\ell,:}\right), \tag{4.19}$$

where N_d is the number of OFDM blocks, and $(\mathbf{G}_t)_{\ell,:}$ is the ℓ-th row.

For 2D compressive sensing, we introduce delay dictionary $\boldsymbol{\Psi}_1$ and AoA dictionary $\boldsymbol{\Psi}_2$ (both overcomplete DFT matrices):

$$\mathbf{Y}_t \approx \boldsymbol{\Psi}_1 \cdot \mathbf{D}'_t \cdot \boldsymbol{\Psi}_2^T, \tag{4.20}$$

where $\mathbf{D}'_t$ is an $N_1 \times N_2$ sparse matrix (N_1, N_2: dictionary sizes). Solved via 2D-OMP, non-zero elements (i, j) correspond to τ_i and ϕ_j.

Doppler estimation:

$$f_{D,\ell} \approx \frac{1}{2\pi T_s} \angle \left(\hat{\mathbf{D}}_t \hat{\mathbf{D}}^*_{t+1} \right)_{\ell,\ell}. \tag{4.21}$$

where $\hat{\mathbf{D}}_t$ is the estimated $\mathbf{D}'_t$. Accuracy improves through multi-DMRS signal averaging. Key advantage: resolves multipath with identical delays but different AoAs, outperforming 1D CS.

For 3D compressive sensing, we combine multiple ISAC signals $\mathbf{Y}_{n,t}$ into 3D tensor $\mathcal{Y}$:

$$\mathcal{Y} \approx \boldsymbol{\Psi}_1 \circ \boldsymbol{\Psi}_2 \circ \boldsymbol{\Psi}_3 \cdot \mathcal{X}$$

where "$\circ$" denotes Kronecker product, and $\mathcal{X}$ is a 3D sparse coefficient tensor. Solved via Tensor-OMP, non-zero indices (i, j, k) correspond to $\tau_i, \phi_j, f_{D,k}$.

Doppler dictionary properties: Due to small f_D, $\boldsymbol{\Psi}_3$ uses highly overcomplete DFT matrices (10–20× actual bandwidth) to capture fine frequency variations. However, computational complexity is extreme (maintaining three dictionaries + tensor operations), and Doppler estimation suffers from sensitivity to minor frequency shifts.

By stacking signals from the spatial domain and Doppler-frequency domain, two Multiple Measurement Vector (MMV) CS problems can be formed. Through MMV-CS amplitude estimation, AOA and Doppler frequencies can be estimated effectively [9, 10].

Overall, grid-based CS algorithms show great promise in sensing parameter estimation. However, quantization error poses a significant challenge, particularly for continuous-valued parameters, often resulting in a mismatch between assumed and actual dictionaries, leading to degraded performance—especially problematic when the number of unknown variables is large [11].

A major advantage of CS for perceptual parameter estimation is that it does not require consecutive samples. However, typical compressive sensing techniques use a grid quantization dictionary, and the quantization produces errors when the original parameters have continuous values.

4.4.4 Deep Learning

Complex ISAC scenarios—especially challenging indoor and urban outdoor environments with rich data propagation—exhibit nonlinear channel properties that defy conventional modeling [12]. This renders traditional signal processing inadequate for joint communication-sensing optimization. Deep learning methodologies address this gap by learning representations of channel dynamics, environmental interactions, and system uncertainties. Notable implementations include: a DNN-based demodulator with YOLOv5-SORT tracking [13], model-driven block processing networks [14], and dual-path NN architectures for THz-ISAC [15].

In the presence of non-ideal effects such as Doppler shifts and phase noise, Com-Net has achieved superior Bit Error Rate (BER) performance compared to existing DL methods and traditional detection approaches. Extensive simulation results indicate that the proposed DL methods can mutually enhance communication and sensing performance, demonstrating robustness against Doppler effects, phase noise, and multi-target estimation.

4.5 Conclusion

In this chapter, we provided an overview of FD-ISAC, including the system model and the signal processing flow. Specifically, we described the three key steps for realizing FD-ISAC, i.e., SIC, echo signal extraction and radar estimation. For the SIC technique, it can be detailed in the previous chapter. For echo signal separation, we introduced three methods, namely ICA, multi-tap analog domain cancellation, and Doppler-based dynamic target extraction. For radar estimation, we explored four representative techniques, including periodogram methods, subspace spectrum analysis, CS and DL methods, as well as their characteristics and limitations.

References

1. He Y, Zhao H, Shao S (2024) Nonlinear self-interference cancellation in vehicle networks for full-duplex integrated sensing and communication. IEEE Trans Veh Technol 73(9):13980–13985
2. Weng H-L, Chih Y-C, Shen C-A, Fouda M E, Eltawil A (2023) Enabling integrated sensing and communication using in-band full-duplex systems. In: Proceedings of 57th asilomar conference signals system computing, pp 410–414
3. Baquero Barneto C, Riihonen T, Turunen M, Anttila L, Fleischer M, Stadius K, Ryynänen J, Valkama M (2019) Full-duplex OFDM radar with LTE and 5G NR waveforms: challenges, solutions, and measurements. IEEE Trans Microwave Theory Tech 67(10):4042–4054
4. Duan B, Chen C, Pan W, Shen Y, Liu Y, Shao S (2024) Frequency-domain differential interference cancellation for full-duplex OFDM ISAC systems. IEEE Trans Veh Technol 73(6):8615–8631
5. Li J, An S, An J, Zirath H, He ZS (2020) OFDM radar range accuracy enhancement using fractional fourier transformation and phase analysis techniques. IEEE Sensors J 20(2):1011–1018
6. Jiang W, Wei Z, Feng Z, Chen X (2025) Integrated sensing and communication enabled sensing base station: system design, beamforming, interference cancellation and performance analysis. China Commun 22(1):111–127
7. Liu Y, Liao G, Chen Y, Xu J, Yin Y (2020) Super-resolution range and velocity estimations with OFDM integrated radar and communications waveform. IEEE Trans Veh Technol 69(10):11659–11672

8. Ozkaptan CD, Zhu H, Ekici E, Altintas O (2024) A mmWave MIMO joint radar-communication testbed with radar-assisted precoding. IEEE Trans Wireless Commun 23(7):7079–7094
9. Rahman M L, Cui P, Zhang J A, Huang X, Guo Y J, Lu Z (2019) Joint communication and radar sensing in 5G mobile network by compressive sensing. In: Proceedings of 19th international symposium communication information and technology (ISCIT), pp 599–604
10. Rahman ML, Zhang JA, Huang X, Guo YJ, Heath RW (2020) Framework for a perceptive mobile network using joint communication and radar sensing. IEEE Trans Aerosp Electron Syst 56(3):1926–1941
11. Chi Y, Scharf LL, Pezeshki A, Calderbank AR (2011) Sensitivity to basis mismatch in compressed sensing. IEEE Trans Signal Process 59(5):2182–2195
12. De Lima C, Belot D, Berkvens R, Bourdoux A, Dardari D, Guillaud M, Isomursu M, Lohan E-S, Miao Y, Barreto AN, Aziz MRK, Saloranta J, Sanguanpuak T, Sarieddeen H, Seco-Granados G, Suutala J, Svensson T, Valkama M, Van Liempd B, Wymeersch H (2021) Convergent communication, sensing and localization in 6G systems: an overview of technologies, opportunities and challenges. IEEE Access 9:26902–26925
13. Zhang Z, Chang Q, Xing J, Chen L (2023) Deep-learning methods for integrated sensing and communication in vehicular networks. Veh Commun 40:100574
14. Jiang W, Ma D, Wei Z, Feng Z, Zhang P, Peng J (2024) ISAC-NET: model-driven deep learning for integrated passive sensing and communication. IEEE Trans Commun 72(8):4692–4707
15. Wu Y, Lemic F, Han C, Chen Z (2023) Sensing integrated DFT-spread OFDM waveform and deep learning-powered receiver design for terahertz integrated sensing and communication systems. IEEE Trans Commun 71(1):595–610

Chapter 5
Hardware Implementation of FD-ISAC

Abstract Recently, several hardware platforms have been proposed to validate the feasibility of FD-ISAC systems. In the following subsections, we will provide a comprehensive investigation of these hardware implementations. Specifically, we will delve into the hardware realizations of FD SIC techniques, which are crucial for mitigating SI in simultaneous transmission and reception scenarios. FD SIC techniques can be categorized into three domains: antenna-domain, RF-domain and digital-domain, and their respective hardware implementations are described. Moreover, the FD testbeds that integrate these three-domain SIC techniques are also investigated, which can achieve more than 120 dB cancellation performance over 20 MHz bandwidth. Additionally, we will explore the hardware platforms designed for FD sensing, focusing on their ability to simultaneously perform sensing and communication tasks in an efficient and integrated way. Through this investigation, we aim to highlight the latest advancements in the development of hardware solutions for FD-ISAC. However, it should be noted that hardware implementations of FD-ISAC remain in the early stages.

5.1 Hardware Implementation of FD SIC

The hardware implementation of SIC has been widely carried out. As discussed in Sect. 3.3 and illustrated in Fig. 3.5, there are mainly three categories of SIC based on where signal cancellation occurs, i.e., antenna-domain SIC (ASIC), RF-domain SIC (RFSIC), and digital-domain SIC (DSIC). In practical FD systems, the combination of ASIC, RFSIC and DSIC can effectively suppress SI to a level close to the noise floor. The first step is to perform an ASIC scheme using a variety of antenna techniques. After that, we can sequentially employ RFSIC and DSIC to cancel residual SI.

C. Du et al., *Full-Duplex Integrated Sensing and Communication Systems*,
https://doi.org/10.1007/978-981-92-0470-0_5

5.1.1 Hardware Implementation of ASIC

Due to the high power of SI signal in FD terminals, it will overwhelm the signal of interest and potentially exceed the dynamic range of the ADC at the receiver [1, 2]. To mitigate these effects, it is essential to employ ASIC technique as the first stage of defense against SI in FD system. ASIC technology suppresses SI by physically isolating the transmitting and receiving links, producing an equivalent SI-attenuation effect. Moreover, ASIC plays an important role and accounts for a large portion of SIC in existing FD designs[3]. As demonstrated in [4], the total 85 dB of SIC was achieved in the FD design, and 65 dB was attributed to the application of ASIC. In the following, single-antenna and multiple-antenna ASIC techniques will be separately discussed in detail.

(1) Single-antenna ASIC

The transmitting and receiving link of single-antenna FD devices can be separated using isolation components such as circulators [5, 6] and EBD [7, 8], etc. Circulators are a class of nonreciprocal components in which the signal can only travel in one direction. In contrast, EBD employs impedance balancing techniques through a hybrid transformer structure and a tunable balancing network to achieve signal cancellation at the receiver input. Hardware implementations of these two single-antenna ASIC schemes will be discussed in detail in the following part.

- **Circulator:** A three-port circulator can be used in the antenna interface of FD systems to connect the Tx and the Rx to a single antenna while isolating the two from each other. Circulators can be implemented using magnetic technologies based on ferrite materials and external magnetic fields, or non-magnetic approaches employing metamaterials, spatiotemporal modulation, or integrated circuits. In [5], a magnetic circulator combined with a reconfigurable impedance mismatched terminal (IMT) circuit was employed to cancel the primary SI signals leaked from the Tx port to the Rx port by using the inherent secondary SI signals at the circulator, as shown in Fig. 5.1. The proposed design achieved more than 40 dB cancellation at 2.45 GHz over 65 MHz bandwidth. However, magnetic circulators are often bulky, expensive, and difficult to integrate with modern CMOS-compatible platforms. In [6], a non-magnetic, nonreciprocal circulator-receiver architecture based on an N-path filter was proposed, which combined a commutation-based linear periodically timevarying non-magnetic circulator with a down-converting mixer and directly provided the baseband RX signals at its output. The proposed design achieved 40 dB average isolation at 0.61∼0.975 GHz over 20 MHz radio RF bandwidth.

Lessons Learned **:** Magnetic circulators exhibit excellent isolation performance and low insertion loss, but suffer from bulkiness, high cost, and poor compatibility with modern CMOS-compatible platforms. In contrast, non-magnetic circulators are more integration-friendly by eliminating the need for ferrite materials and external

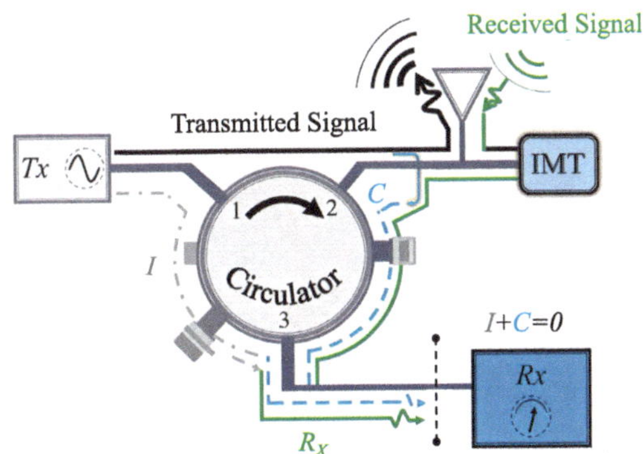

Fig. 5.1 Single-antenna FD system integrated with the magnetic circulator and IMT circuit [5]

magnetic biasing. However, further improvements of the circulator's linearity performance, the TX power handling capability and improved balancing schemes should be considered to provide a wider range of antenna impedance mismatches.

- **EBD:** The circuit of EBD exploits the electrical balancing of signals within a hybrid junction and impedance matching network to achieve high isolation between Tx and Rx port. When all ports of the hybrid junction are properly impedance-matched, an input signal injected at any port will be directed to the two adjacent ports without coupling to the opposite port. EBD can be categorized into passive EBD and active EBD. Passive EBD uses fixed impedance networks with passive components to maintain balance, while active EBD employs tunable impedance and real-time feedback control to adapt to antenna impedance variations. In [7], a passive EBD providing 44 dB/45 dB isolation at 890 MHz/1890 MHz over 20 MHz bandwidth was combined with an RF canceller to increase isolation to 83 dB of combined EBD and RF cancellation. However, the practical antenna impedance severely limits the cancellation bandwidth in the passive balancing network. In [8], a hardware prototype compound EBD combining both passive and active balancing techniques was constructed, as illustrated in Fig. 5.2. The compound EBD achieved higher cancellation performance of 81.5 dB at 1.9 GHz over 80MHz bandwidth. The prototype system exhibited an approximately 15 dB increase in noise floor when employing active EBD compared to the passive EBD; however, the compound EBD showed no additional noise compared with the passive EBD.

Lessons Learned **:** EBD features a wide impedance tuning range, enabling it to support wide-bandwidth signals and overcome the narrow-band limitation of magnetic circulators. Based on a tunable impedance network, it enables real-time adaptation to channel variations, thereby enhancing the robustness of interference suppression in dynamic environments. However, the core working mechanism of EBD relies on real-time impedance matching, and its dynamic impedance matching mechanism necessitates complex real-time tuning algorithms.

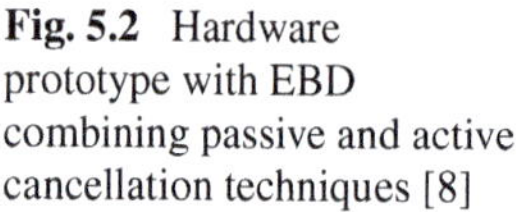

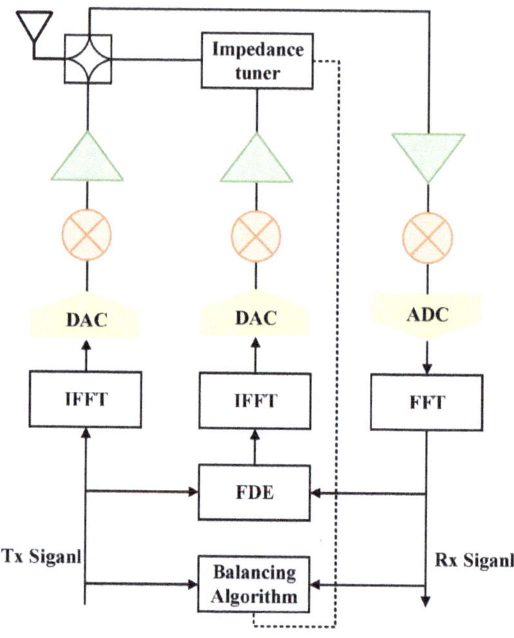

Fig. 5.2 Hardware prototype with EBD combining passive and active cancellation techniques [8]

(2) Multiple-antenna ASIC

Multiple-antenna schemes improve isolation performance by strategically designing and arranging antenna elements to suppress mutual coupling effects. Beamforming [9, 10] and cross-polarization [11–13] technologies are commonly employed to improve the physical separation between the Tx and Rx antennas of multi-antenna FD devices.

- **Beamforming:** Beamforming technique can be applied to FD antenna arrays to manipulate their transmit and receive radiation patterns, leveraging beam directivity to separate SI from the desired signal. In [9], phased-array beamforming technique combined with FD operation was presented to achieve SIC with minimal link budget penalty and with no additional power consumption. Hardware prototype with beamforming can be seen in Fig. 5.3a. The design can provide 50 dB ASIC for +8.7 dBm OFDM-like signal at 0.73 GHz over 5 MHz bandwidth. In [10], a two-stage multichannel SIC system by using transmission and receive beamforming sequentially was presented, as shown in Fig. 5.3b. The experimental result showed that the proposed method achieved 27 dB ASIC cancellation at 4.5 GHz over 20 MHz bandwidth with +12 dBm transmit power.

Lessons Learned **:** Beamforming can suppress SI signals in the antenna domain through the directional radiation characteristics of antenna arrays, and can be cascaded with DSIC to improve the overall performance without relying on additional hardware resources. However, it imposes extremely high requirements on the cal-

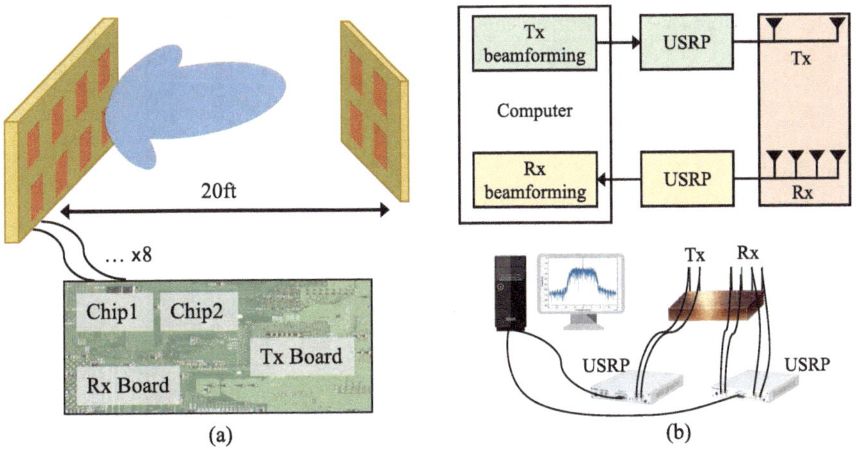

Fig. 5.3 **a** Hardware prototype with beamforming[9]. **b** System architecture and hardware prototype with beamforming [10]

ibration accuracy of antenna arrays and the real-time availability of channel state information.

- **Cross-polarization:** Cross-polarization-based cancellation is an additional mechanism for isolating the Tx and Rx antennas electromagnetically. Polarization decoupling operates the transmit and receive antennas with orthogonal horizontal and vertical polarizations. In [11], an investigation of the wideband SI channel characteristics of a 2×2 MIMO FD transceiver using cross-polarized antennas was presented. In their design, cross-polarized antennas were considered for compact size and high isolation by using comb-corrugated Vivaldi ridges integrated in a circular horn. The cross-polarized antennas can achieve $>$ 45 dB cancellation at 2.45 GHz over 500 MHz bandwidth with average +5 dBm transmitted power. In [13], two cross-polarized microstrip patch antenna systems were proposed, which were based on a single radiating element with 180° ring hybrid coupler for differential feeding. It illustrates the fabricated prototypes of two cross-polarized microstrip patch antennas. One of the implemented single-layer patch antennas provided more than 67 dB isolation between Tx and Rx ports at 2.4 GHz while the second fabricated antenna with slot coupled Tx port along with differential feeding for Rx operation provides more than 90 dB RF isolation at 2.41 GHz. The vertical dimensions of both antennas can be decreased by reducing the distance between radiating patch and coupler; however, the additional mutual coupling caused by close proximity may degrade the interport isolation.

Lessons Learned **:** Cross-polarization uses orthogonally polarized antennas to suppress co-polarized SI signals without relying on complex signal processing, which features relatively simple hardware implementation. However, this approach exhibits high dependency on polarization purity and is susceptible to environmental scatter-

Table 5.1 Performance summary and comparison of ASIC technologies

SIC method	Work	Frequency (GHz)	Bandwidth (MHz)	Transmitted power (dbm)	Measured cancellation (DB)
Circulator	[5]	2.45	65	18	40
Circulator	[6]	0.61~0.975	20	+8	40
EBD	[7]	890 MHz/1890 MHz	20	+10	44/45
EBD	[8]	1.9	80	–	81.5
Beamforming	[9]	0.73	5	+8.7	50
Beamforming	[10]	4.5	20	+12	27
Cross-polarization	[11]	2.45	500	+5	45
Cross-polarization	[13]	2.4	50	–	67

ing and polarization rotation, which cause polarization leakage and thus degrade isolation performance.

In summary, Table 5.1 provides a performance comparison of several published ASIC cancellers.

5.1.2 Hardware Implementation of RFSIC

RFSIC is commonly employed to mitigate SI at the RF front end, preventing saturation of the analog-to-digital converters (ADC). By performing cancellation at the RF stage, RFSIC alleviates the burden on the ADC, allowing for enhanced dynamic range and improved signal fidelity. The objective of RFSIC is to reconstruct an accurate copy of the SI waveform and subtract it from the received signal in the RF domain, thereby effectively reducing the SI. RFSIC techniques can be classified into two major types, i.e., the multi-tap and the digital-assisted approaches.

(1) Multi-tap approach

The multi-tap approach couples the RF transmitted signal and guides it through phase-amplitude adjustment circuits to rebuild the SI signal as well as the transmitter noise. In particular, multi-taps RF canceller design has been proved to be the most effective way to suppress multipath SI [14–17]. However, the reconstruction process of cancellation signals is a typical multi-dimensional optimization process whose optimization complexity grows exponentially with the number of TAPs. Therefore, designing a fast and stable convergence optimization method is the key to achieving real-time and high-precision RFSIC.

The Gradient Descent (GD) mechanism is recognized as an effective approach to solve the multi-dimensional optimization problems. Its efficacy stems from its

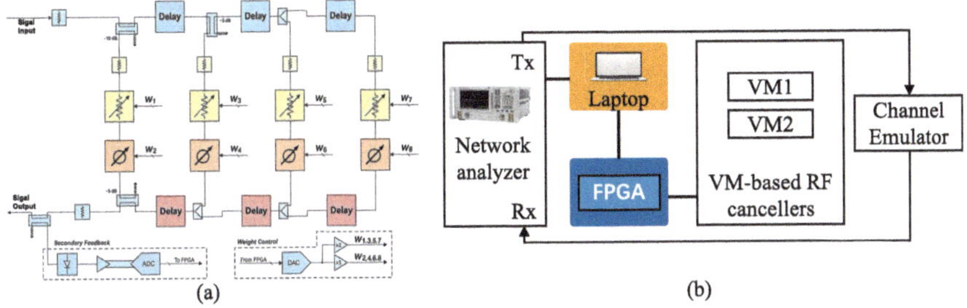

Fig. 5.4 **a** The four-tap RF canceller prototype [14]. **b** The two-tap RF canceller prototype[17]

ability to efficiently navigate complex parameter spaces by following the steepest descent direction of the objective function. In [14], a 4-TAP RF canceller employing a modified dithered linear search (DLS) algorithm was designed, as shown in Fig. 5.4a. Although the design exhibited non-negligible insertion loss, it was intentionally implemented as a fully passive structure to help suppress the uncorrelated noise and nonlinearities of the canceller circuit. The approach can achieve 22dB RF cancellation for +20dBm OFDM signal at 2.45 GHz over 20 MHz bandwidth. However, it requires 500 iterations and 500 μs convergence time.

Moreover, in [16], an 8-TAP RF canceller with an iterative algorithm based on GD was proposed, which can achieve 38 RF cancellation at 2.4 GHz over 20 MHz bandwidth with +10 dBm average transmitted power, requiring only 30 iterations to converge. Although the proposed algorithm can theoretically converge to the global optimum, its practical performance is still constrained by the resolution of the DAC used for controlling the amplifiers and phase shifters, and non-ideal properties of analog devices.

Furthermore, in [17], a 2-TAP RF canceller with an adaptive tuning approach called ATEGrad was proposed, as seen in Fig. 5.4b. The algorithm dynamically adjusted its learning rate by utilizing a target-error parameter and a gradient estimation that is gradually forgotten to maximize cancellation with a minimal convergence time. In their design, it can achieve 35.2 dB RF cancellation at 2.5 GHz over 20 MHz bandwidth.

Lessons Learned **:** The multi-tap approach facilitates efficient suppression of transmitter-induced noise, thereby enhancing SIC performance. However, this technique incurs higher hardware complexity, as it requires multiple adjustable amplitude-phase control paths with independent delay lines.

(2) Digital-assisted approach

The digital-assisted approach extracts the transmitted signal in its digital form and subsequently routes it through an auxiliary transmit chain to reconstruct the RF SI signal[1]. Different from multi-tap approach, where the number of taps and the resolution of phase and amplitude adjustments are often constrained by hardware limitations, digital assisted-tapping can dynamically adapt to varying channel conditions and SI profiles with minimal hardware modifications.

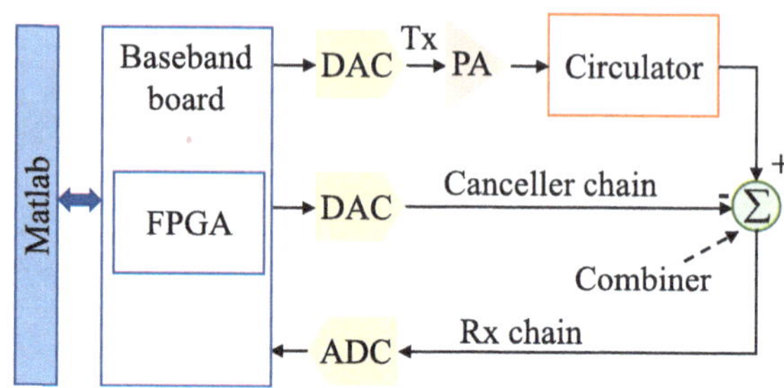

Fig. 5.5 Block diagram and photograph of the testbed designed for digital-assisted RFSIC verifications [18]

In [18], a digital-assisted RF SIC architecture was proposed, as shown in Fig. 5.5, which included an extra observation chain to capture the nonlinear behavior of the transmitter, along with a nonlinear model to reproduce its nonlinearity to further mitigate the nonlinear components of the SI. Their canceller demonstrated the capability to achieve 41 dB RF cancellation on a 20 MHz LTE signal with +31 dBm average transmitted power, while linear methods can only achieve 32 dB RF cancellation.

Moreover, in [19], a digital-tapping canceller with a novel block based on least mean squares(LMS) algorithm was considered, along with a closed-loop parameter learning approach based on the decorrelation learning rule, which enabled efficient estimation of the coefficients of the nonlinear cancellation filter. The proposed design can achieve 54dB RF cancellation at 2.14 GHz over 20 MHz bandwidth with average +30 dBm transmit power.

Furthermore, [20] proposed a new filter LMS algorithm for digital-tapping canceller, along with a novel orthogonalization procedure for nonlinear basis functions. The design can provide 48 dB RF cancellation at 2.4 GHz over 18MHz bandwidth with average +8 dBm transmitted power, suppressing the residual SI to within 2 dB above the noise floor. However, it is worth noting that the cancellation of non-ideal effects such as phase noise has not been carried out in their work. In the design of the digitally-assisted RFSIC, it is crucial to take into account the effects of the auxiliary chains. SI cancellation using an auxiliary Tx may be constrained by phase noise differences between the main and the auxiliary Tx chain, as demonstrated in prior work [21]. Thus, the nonlinear distortions introduced by the auxiliary Tx chain may have to be considered to further improve the cancellation performance.

Lessons Learned **:** In digital-assisted approach, delay and attenuation adjustments as well as arbitrary number of TAPs can be seamlessly implemented in the digital domain, offering greater flexibility and precision in controlling the cancellation process. However, this approach faces limitations in canceling transmitter noise effectively because the random noise components of the primary and auxiliary transmit chains are statistically independent. As a result, the auxiliary signal cannot accurately replicate and subtract the transmitter noise, leaving residual interference in the system.

In summary, Table 5.2 describes the performance of several published RFSIC cancellers.

Table 5.2 Performance Summary and Comparison of RFSIC technologies

SIC method	Work	Frequency (GHz)	Bandwidth (MHz)	Transmitted power (dBm)	Measured cancellation (dB)
Multi-tap approach	[14]	2.45	20	+20	22
Multi-tap approach	[16]	2.45	20	+10	38
Multi-tap approach	[17]	2.5	20	–	35.2
Digital-assisted approach	[18]	–	20	+31	41
Digital-assisted approach	[19]	2.14	20	+30	54
Digital-assisted approach	[20]	2.4	18	+8	48

5.1.3 Hardware Implementation of DSIC

The last stage of SI cancellation is performed in the digital-domain, aiming to suppress the remaining SI from the received signal, and in the best case, to reduce it below the noise floor. Basically, the estimated residual SI is subtracted from the received signal in the digital domain [4]. Working in the digital-domain offers significant implementation advantages that sophisticated processing becomes relatively easy through the use of advanced digital signal processing (DSP) techniques [22]. In general, DSIC techniques include two types: channel estimation and adaptive filter.

(1) Channel estimation

For the channel estimation method, firstly, the near-end transmit signal is coupled as the feedback signal, and the channel estimation is performed by some algorithms between the feedback signal and the received signal. After obtaining the SI channel estimation weights, the feedback signal will be multiplied with the estimation weights to reconstruct the SI signal, and subtract the copy of SI signal from the received signal to suppression SI.

The least squares (LS) estimator has been commonly adopted for acquiring linear and nonlinear SI channel parameters. For instance, in [23], a digital canceller was proposed to jointly suppress both linear and non-linear SI, in which an efficient LS parameter estimator was designed to cancel linear components. Experimental validation demonstrated that the proposed DSIC scheme achieved approximately 38 dB cancellation at 2.45 GHz over 20 MHz bandwidth with +20 dBm transmitted power. Moreover, in [24], three alternative monostatic antennas configurations exhibiting

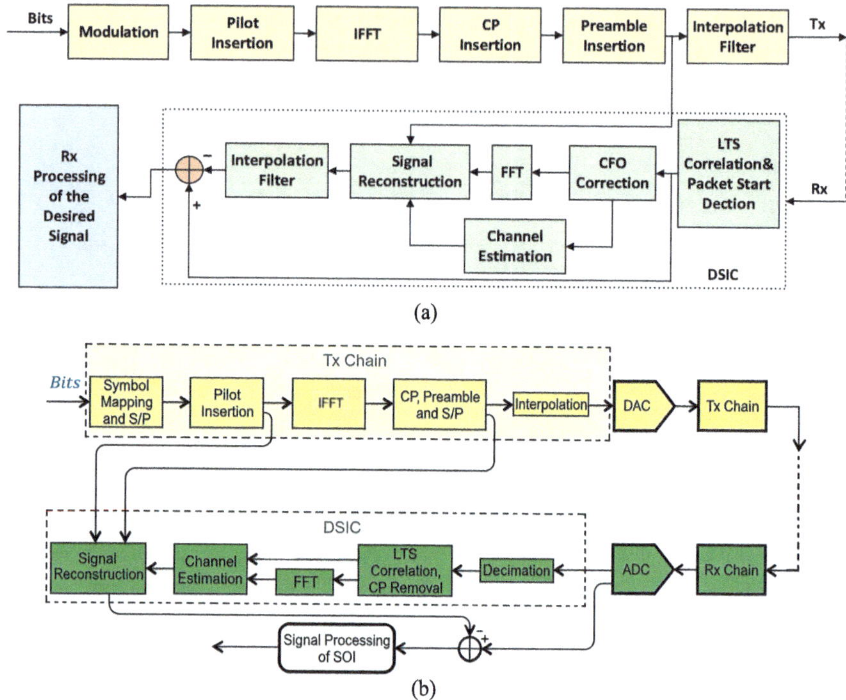

Fig. 5.6 **a** Time-domain LS estimation technique[24]. **b** Frequency-domain LS estimation technique [25]

high levels of ASIC were integrated with a linear DSIC algorithm based on the LS method. A time-domain LS estimation technique was utilized to estimate the SI channel parameters, as shown in Fig. 5.6a. The DSIC scheme was implemented on a WARP v3 platform, which can achieve 28 dB DSIC at 2.4 GHz over 20 MHz bandwidth with average +15 dBm transmitted power. Furthermore, in [25], a frequency-domain LS estimation method was proposed to cancel linear SI in OFDM system, as shown in Fig. 5.6b. The design maintained effective cancellation performance and reduced the complexity cost of 61% compared to conventional time-domain LS estimation, which can achieve 26.6 dB DSIC at 2.4 GHz over 20 MHz bandwidth with average +4.3 dBm transmitted power.

Lessons Learned **:** Channel estimation method such as the LS estimator offers a closed-form solution for estimating the SI channel parameters, which makes it computationally efficient and straightforward to implement. However, a key limitation of the LS estimator lies in its inability to adapt to time-varying or rapidly changing channel conditions. This lack of adaptability restricts its effectiveness in dynamic wireless environments.

(2) Adaptive filter

Adaptive filter is also recognized as an effective means to rebuild SI in digital-domain, which can be well-suited for hardware implementation on a FPGA. The core principle of adaptive filter is that the filter can dynamically adjust its parameters in real time by sensing the change of the environment, so that it can obtain the optimal DSIC performance.

In [26], a new orthogonalization method for nonlinear basis functions was proposed for DSIC, where the LMS algorithm was adopted for parameter estimation of parallel Hammerstein signal model. Experimental results indicated that approximately 46 dB of DSIC can be achieved at 2.46 GHz over 20 MHz bandwidth with average +13.3 dBm transmitted power. However, 10^4 iterations are required to reach convergence for their algorithm, which poses significant limitations for real-time applications and is generally considered impractical for latency-sensitive FD systems. In [27], a 28-taps adaptive filter was designed in the digital-domain, utilizing LMS algorithm to update the tap coefficients in real-time. It achieved 30 dB of DSIC at 1.74 GHz over 40 MHz bandwidth with average −5 dBm transmitted power. However, wideband operation, a rapidly changing RF channel, and a relatively long delay spread pose significant challenges for the real-time implementation of an FD system. In [28], an optimized LMS adaptive filter architecture that employed subband decomposition and parallel processing techniques was proposed to improve the wideband SI cancellation performance in a representative 5G environment, which provided 50.5 dB cancellation at 1.5 GHz over 1 GHz bandwidth (Fig. 5.7).

***Lessons Learned* :** Adaptive filtering algorithms play a crucial role in enabling rapid and continuous tracking of time-varying SI channels, thereby enhancing the robustness of DSIC in dynamic environments. However, their performance is sensitive to initialization conditions. Inappropriate initialization may cause the algorithm to converge to a local optimum rather than the desired global minimum, ultimately resulting in lower cancellation performance.

In summary, Table 5.3 provides the performance comparison of DSIC cancellers presented in this section.

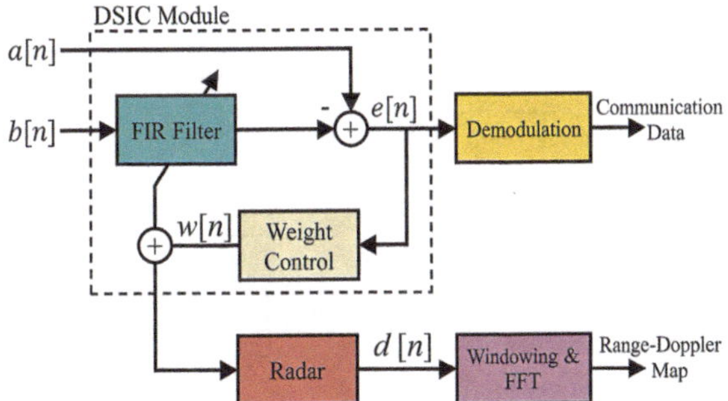

Fig. 5.7 DSIC based on adaptive filter[27]

Table 5.3 Performance summary and comparison of DSIC technologies

SIC Method	Work	Frequency (GHz)	Bandwidth (MHz)	Transmit power (dBm)	Measured cancellation (dB)
Channel estimation	[23]	2.45	20	+20	38
Channel estimation	[24]	2.4	20	+15	28
Channel estimation	[25]	2.4	20	+4.3	26.6
Adaptive filter	[26]	2.46	20	+13.3	46
Adaptive filter	[27]	1.74	40	−5	30
Adaptive filter	[28]	1.5	1	–	50.5

5.1.4 Hardware Implementation of Three-Domain SIC

Combining with three-domain SIC techniques, several works set up FD testbed to validate system performance under practical conditions. Typical hardware implementations of three-domain SIC are presented in Figs. 5.8, 5.9 and 5.10. In [23], Stanford university proposed an FD testbed based on WARP, which can achieve a total 110 dB cancellation (15 dB in antenna-domain, 57 dB in RF-domain and 38 dB in digital-domain) at 2.4 GHz over 20 MHz bandwidth with average +20 dBm transmitted power and OFDM 64QAM modulated SI signal. In [29], Yonsei University presented an FD radio system for 5G wireless networks, which can achieve a total 103 dB cancellation (42 dB in antenna-domain, 18 dB in RF-domain and 43

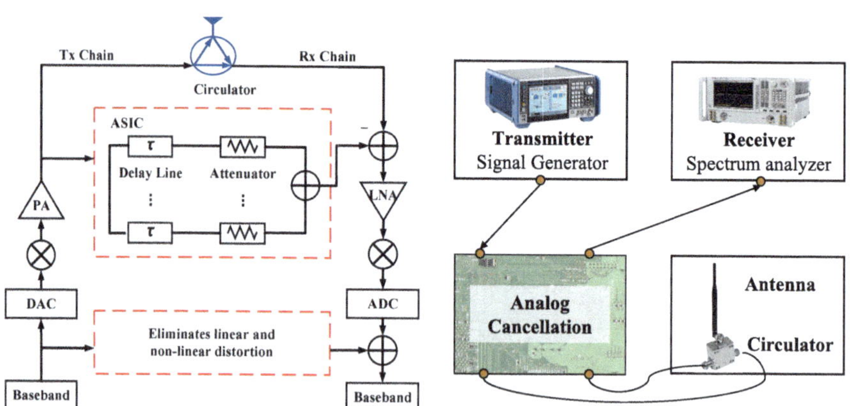

Fig. 5.8 Block diagram of the FD radio [23].

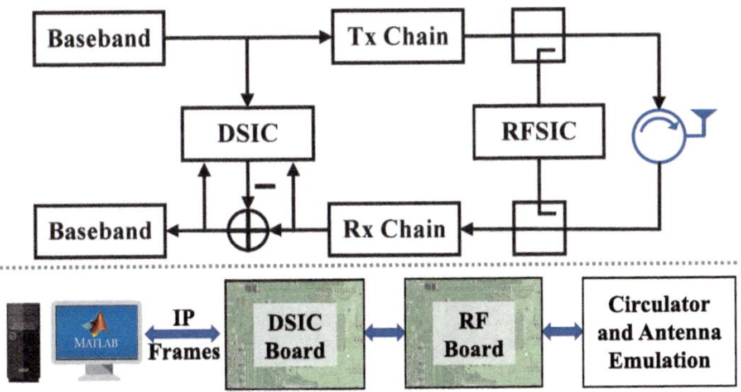

Fig. 5.9 Block diagram of the FD transceiver [30].

dB in digital-domain) at 2.52 GHz, over 20 MHz bandwidth with average +20 dBm transmitted power.

Moreover, in [30], Intel proposed an FD testbed based on AD9361 from Analog Devices, Inc. (ADI), which can achieve a total 91 dB cancellation (21 dB in antenna-domain, 36dB in RF-domain and 34 dB in digital-domain) at 900 MHz, over 20 MHz bandwidth with average +20dBm transmitted power.

Furthermore, in [31], Tampere university proposed an FD relay testbed, which can achieve a total 115.3 dB cancellation (75.5 dB in antenna-domain, 16.9 dB in RF-domain and 22.9 dB in digital-domain) at 2.56 GHz, over 20 MHz bandwidth with average +29 dBm transmitted power. However, the Hammerstein model in [31] cannot handle the other impairments intrinsic to direct-conversion transmitters such as I/Q mismatch, LO leakage, and baseband nonlinearities. In [32], Tampere University introduced an FD testbed based on a software-defined radio (SDR) platform, which presented a novel, low complexity SI canceller algorithm based on an extended Hammerstein model containing the most significant additional transmitter impairments. The system can achieve a total SI suppression of 103 dB(40 dB in antenna-domain, 26 dB in RF-domain and 37 dB in digital-domain) at 2.4 GHz, over 10 MHz bandwidth with average +20 dBm transmitted power. In addition, Ref. [33] proposed an FD base station, which can achieve a total 122 dB cancellation (70 dB in antenna-domain, 12 dB in RF-domain and 40 dB in digital-domain) at 2.4 GHz and 3.5 GHz, over 20 MHz bandwidth with average +32 dBm transmitted power.

To further improve the SIC performance, we developed a new FD testbed, which consists of three parts: vertical/horizontal polarization antennas, RF SIC board and digital SIC board, as illustrated in Fig. 5.11a. The polarization-based transceiver antennas can provide >40 dB antenna cancellation with > 15 dB antenna gain. A 4-TAP RF canceller was designed and a variable step approaching (VSA) algorithm was developed for adaptively tuning the weights of each TAP, which detailed in [15]. The digital SIC board mainly consists of an AD9371 transceiver from ADI and an FPGA (Virtex-7) from Xilinx, and a nonlinear digital SIC algorithm combined with parallel Normalized Least Mean Square (NLMS)-based parameter learning is also

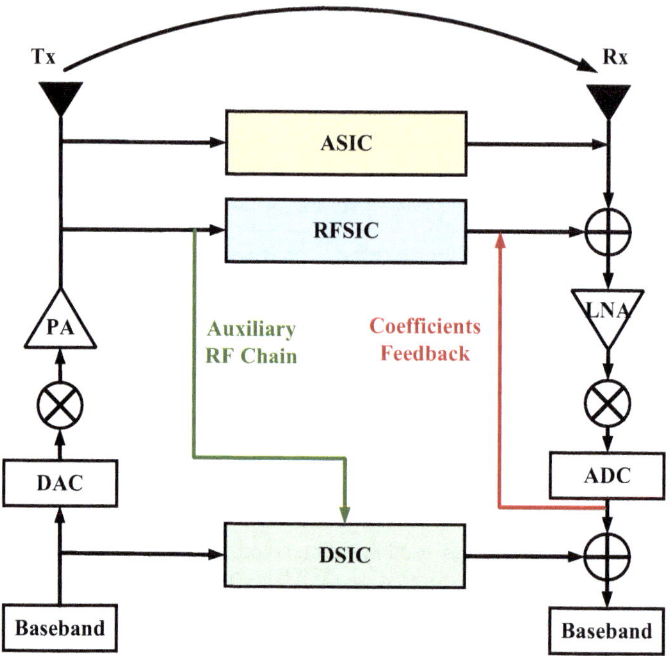

Fig. 5.10 Block diagram of the FD prototype [33]

developed, as detailed in [34]. The measurement scenario is depicted in Fig. 5.11b, where two FD nodes are positioned on the rooftops of two separate buildings. In our experimental setup, we utilize AD9371 to generate 1.96GHz 64QAM transmit signal and subsequently amplified to +30 dBm through a power amplifier.

The measured power spectral densities of the received SI before and after performing the RFSIC and digital SIC are shown in Fig. 5.11c, d. Under 40 dB antenna isolation, the received SI power is suppressed to −10 dBm. By reducing to −54 dBm after performing RFSIC and further reducing to -96dBm after performing DSIC, it achieves a total SI cancellation of 126 dB over 20 MHz bandwidth. Table 5.4 presents a comparative summary of SIC performance between these published prototypes and our proposed work. Our proposed FD device demonstrates higher SIC performance. Due to +30dBm transmitted power, >15 dB antenna gain and 126 dB overall SIC capability, our FD device can achieve >5 km wireless communication.

5.2 Hardware Implementation of FD Sensing

Leveraging high-performance SIC techniques, the wireless device is capable of simultaneously transceiving on the same frequency, therefore several studies focus

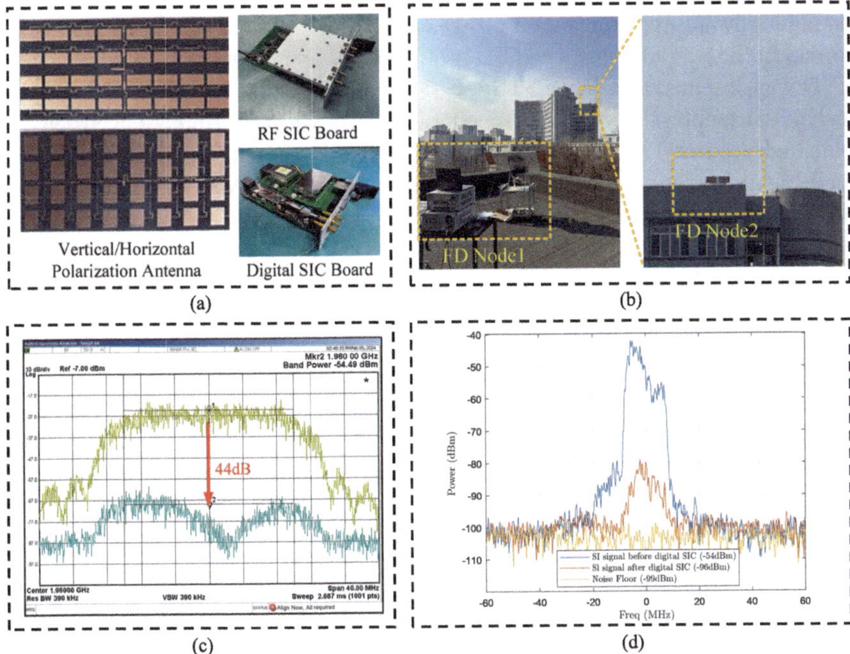

Fig. 5.11 **a** Photograph of experimental setup for SIC and communication. **b** Photograph of measurement scenario. **c** Power spectral density of SI before and after RF SIC. **d** Power spectral density of SI before and after digital SIC

Table 5.4 Performance Summary and Comparison of three-domain FD technologies

	[23]	[29]	[30]	[31]	[32]	[33]	Our proposed work
Frequency	2.45 GHz	2.52 GHz	0.9GHz	2.56 GHz	2.4GHz	2.4 GHz and 3.5GHz	1.96GHz
Modulation	64QAM	4/16/64QAM	QPSK	BPSK	OFDM	16QAM	64QAM
Bandwidth	20 MHz	20 MHz	20 MHz	20 MHz	10 MHz	20 MHz	20 MHz
TX power	+20 dBm	+20 dBm	+20 dBm	+29 dBm	+20 dBm	+32 dBm	+30 dBm
Antenna cancellation	15 dB	42 dB	21dB	75.5 dB	40 dB	70 dB	40 dB
RF cancellation	57 dB	18 dB	36 dB	16.9dB	26 dB	12 dB	44 dB
Digital cancellation	38 dB	43 dB	34 dB	22.9 dB	37 dB	40 dB	42 dB
Total cancellation	110 dB	103 dB	91 dB	115.3 dB	103 dB	122 dB	126 dB
Above noise floor	1 dB	–	–	1.7 dB	3.5 dB	3 dB	3 dB

on the hardware implementation of FD sensing based on the shared OFDM waveforms. In [35], a radar-enhanced FD prototype consisting of a SDR module and an EBD was designed to achieve 55 dB Tx-Rx isolation in the 1.74 GHz RF band. For an EBD SI suppression of 45 dB, speeds range from 200 to 800 mm/s can be detected with high accuracy. However, the system does not support ranging capability. In [36], an FD system prototype was evaluated, as seen in Fig. 5.12, which consisted of both RF and digital SI canceller modules, enabling > 85 dB Tx-Rx isolation. The experimental result showed that the proposed system can accurately measure the velocity of mobile objects across a range of speeds from 0.2 to 1 m/s, while simultaneously serving as a node to perform in-band bidirectional communication.

In [37], an FD OFDM Radar was proposed, combining with three-domain SIC and frequency-domain radar processing at 2.4 GHz. In the measurement scenario, three vehicles labeled A, B, and C are driving along a road toward the OFDM radar system at distances of $50 \sim 110$ m and the speed of ≤ 50 km/h. With the support of overall 100 dB SIC, their ISAC device is clearly able to sense all three moving vehicles.

In [38], the sensing performance of two FD ISAC systems at different operating frequencies were evaluated through RF measurements. In the first case, a similar vehicular sensing scenario as in [37] using an FD OFDM-based radar at the 2.4 GHz band was presented. This case represented a typical example of downlink sensing operated by a BS. In this experiment, a single horn antenna is shared between the TX and RX through a circulator. In addition, both analog and digital SIC schemes were incorporated, providing an overall isolation of approximately 100 dB. The considered ISAC system was tested in a suburban vehicular traffic scenario with four moving cars using a 5G NR waveform with 40 MHz bandwidth, which provides a distance resolution of about 4 m. In the second case, the indoor uplink mapping results was presented when a measurement setup at 28 GHz. In this measurement campaign, a

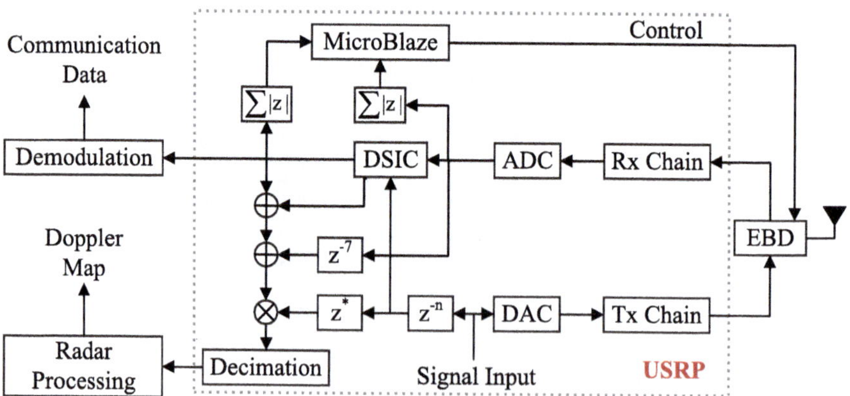

Fig. 5.12 Block diagram of the prototyped FD system[36].

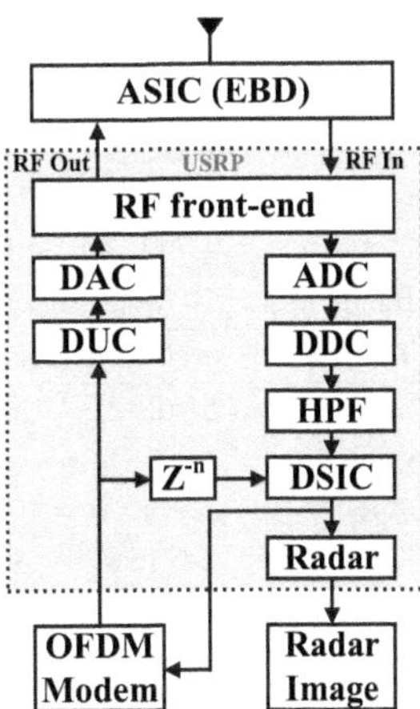

Fig. 5.13 Functional block diagram of the prototyped FD system [27]

5G NR waveform with 400 MHz bandwidth was utilized, which provided much finer resolution of about 0.4m compared to the previous setup.

Moreover, in [27], an 1.74Gz FD single carrier Radar was proposed, which can be assigned to an already standardized framework. The system benefits from EBD module, which provides primary Tx-Rx isolation. This merit is achieved by employing a hybrid transformer coupled with a balance network that emulates the antenna impedance. The dithered linear search algorithm [39] was adopted to tune the EBD, aiming for maximum Tx-Rx isolation. Besides, a DSIC block was deployed to suppress the residual SI signal. At the core, the DSIC module consisted of 28-tap FIR filter and a real-time LMS optimizer. In their design, an overall 80 dB SIC was achieved and the moving targets within a range of $0.5 \sim 4$ m at the speed of ≤ 0.5 m/s can be sensed.

Furthermore, in [40], a 28 GHz FD OFDM Radar was proposed, with static targets highlighted (bicycles near some metallic structures). The system can sense and map an outdoor environment within 20m by using standard-compliant 5G NR waveform with 400MHz channel bandwidth (Fig. 5.13).

Table 5.5 provides a performance summary and comparison of hardware implementation in FD sensing.

In summary, the performance of three-domain SIC has exceeded 120dB with +30dBm transmitted power over 20 MHz bandwidth. However, the hardware implementations of FD-ISAC is still in the early stages. Challenges such as non-ideal analog cancellation circuits, stringent synchronization requirements, and the joint design of sensing and communication functionalities significantly hinder the practical realization of FD-ISAC. Therefore, further advancements in hardware architecture, co-design methodologies, and interference management are essential to bridge the gap between theoretical capabilities and real-world deployment.

Table 5.5 Performance Summary and Comparison of hardware implementation in FD sensing

	[35]	[36]	[27]	[37]	[38]	[40]
SIC technologies	ASIC	ASIC+DSIC	RF SIC+DSIC	ASIC+RF SIC+DSIC	ASIC+RF SIC+DSIC	–
Frequency	1.74 GHz	1.74 GHz	1.74 GHz	2.4 GHz	2.4 GHz	28 GHz
Bandwidth	10 MHz	10 MHz	40 MHz	40 MHz	40 MHz	400 MHz
TX power	+20 dBm	+20 dBm	−5 dBm	+20 dBm	+30 dBm	+30 dBm
SIC	55 dB	85 dB	71 dB	100 dB	100 dB	–
Detection range	–	3.8 m	4 m	110 m	–	20 m
Detection velocity range	2–8 m/s	0.2–1 m/s	0.5 m/s	12 m/s	10 m/s	–

5.3 Conclusion

This chapter has provided an overview of effective SIC techniques implemented in practical FD terminals, and practically achievable performance of each technique. It includes major sections that focus on SIC techniques within the antenna, RF and digital domains as well as dedicated sections on hardware implementations for both FD SIC and FD sensing. Leveraging the aforementioned SIC techniques and testbeds, experimental studies have demonstrated the feasibility and substantial potential of FD wireless systems. However, the hardware implementations of FD-ISAC are still in the early stages. Despite these advancements, a lot of challenges still need to be tackled before FD technology can be widely adopted. More attention from the industry should also be given to develop small-size and low-cost FD solutions.

References

1. Duarte M, Dick C, Sabharwal A (2012) Experiment-driven characterization of full-duplex wireless systems. IEEE Trans Wirel Commun 11(12):4296–4307
2. Kolodziej KE, Perry BT, Herd JS (2019) In-band full-duplex technology: techniques and systems survey. IEEE Trans Microw Theory Techn 67(7):3025–3041
3. Everett E, Sahai A, Sabharwal A (2014) Passive self-interference suppression for full-duplex infrastructure nodes. IEEE Trans Wirel Commun 13(2):680–694
4. Duarte M, Sabharwal A, Aggarwal V, Jana R, Ramakrishnan KK, Rice CW, Shankaranarayanan NK (2014) Design and characterization of a full-duplex multiantenna system for WiFi networks. IEEE Trans Veh Technol 63(3):1160–1177
5. Khaledian S, Farzami F, Smida B, Erricolo D (2018) Inherent self-interference cancellation for in-band full-duplex single-antenna systems. IEEE Trans Microw Theory Techn 66(6):2842–2850

6. Reiskarimian N, Dastjerdi MB, Zhou J, Krishnaswamy H (2018) Analysis and design of commutation-based circulator-receivers for integrated full-duplex wireless. IEEE J Solid-State Circuits 53(8):2190–2201
7. Laughlin L, Zhang C, Beach MA, Morris KA, Haine J (2015) A widely tunable full duplex transceiver combining electrical balance isolation and active analog cancellation. In: Proceedings of IEEE 81st vehicular technology conference (VTC Spring), pp 1–5
8. Laughlin L, Zhang C, Beach MA, Morris KA, Haine JL (2016) Passive and active electrical balance duplexers. IEEE Trans Circuits Syst II Express Briefs 63(1):94–98
9. Dastjerdi MB, Reiskarimian N, Chen T, Zussman G, Krishnaswamy H (2018) Full duplex circulator-receiver phased array employing self-interference cancellation via beamforming. In: Proceedings of IEEE radio frequency integrated circuits symposium (RFIC), pp 108–111
10. Lei L, Saba N, Razul SG (2019) A multichannel self-interference cancellation prototyping system. In: Proceedings of IEEE 2nd 5G world forum (5GWF), pp 427–432
11. Chen F, Morawski R, Le-Ngoc T (2018) Self-interference channel characterization for wideband 2 × 2 MIMO full-duplex transceivers using dual-polarized antennas. IEEE Trans Antennas Propag 66(4):1967–1976
12. Ta SX, Nguyen-Trong N, Nguyen VC, Nguyen KK, Dao-Ngoc C (2021) Broadband dual-polarized antenna using metasurface for full-duplex applications. IEEE Antennas Wirel Propag Lett 20(2):254–258
13. Nawaz H, Tekin I (2017) Dual-polarized, differential fed microstrip patch antennas with very high interport isolation for full-duplex communication. IEEE Trans Antennas Propag 65(12):7355–7360
14. Kolodziej KE, McMichael JG, Perry BT (2016) Multitap RF canceller for in-band full-duplex wireless communications. IEEE Trans Wirel Commun 15(6):4321–4334
15. Zhang H, Xing Z, Du C, Zhang Z (2024) A new tuning algorithm based on variable step approaching for RF self-interference cancellation. IEEE Wirel Commun Lett 13(6):1780–1784
16. Zhang L, Ma M, Jiao B (2019) Design and implementation of adaptive multi-tap analog interference canceller. IEEE Trans Wirel Commun 18(3):1698–1706
17. Kolodziej KE, Cookson AU, Perry BT (2021) Adaptive learning rate tuning algorithm for RF self-interference cancellation. IEEE Trans Microw Theory Techn 69(3):1740–1751
18. Liu Y, Quan X, Pan W, Tang Y (2017) Digitally assisted analog interference cancellation for in-band full-duplex radios. IEEE Commun Lett 21(5):1079–1082
19. Kiayani A, Waheed MZ, Anttila L, Abdelaziz M, Korpi D, Syrjälä V, Kosunen M, Stadius K, Ryynänen J, Valkama M (2018) Adaptive nonlinear RF cancellation for improved isolation in simultaneous transmit-receive systems. IEEE Trans Microw Theory Techn 66(5):2299–2312
20. Soriano-Irigaray FJ, Fernandez-Prat JS, Lopez-Martinez FJ, Martos-Naya E, Cobos-Morales O, Entrambasaguas JT (2018) Adaptive self-interference cancellation for full duplex radio: analytical model and experimental validation. IEEE Access 6:65018–65026
21. Sahai A, Patel G, Dick C, Sabharwal A (2013) On the impact of phase noise on active cancelation in wireless full-duplex. IEEE Trans Veh Technol 62(9):4494–4510
22. Liu G, Yu FR, Ji H, Leung VCM, Li X (2015) In-band full-duplex relaying: a survey, research issues and challenges. IEEE Commun Surv Tutorials 17(2):500–524
23. Bharadia D, McMilin E, Katti S (2013) Full duplex radios. SIGCOMM Comput Commun Rev 43(4):375–386
24. Erdem M, Ayar H, Nawaz H, Gurbuz O, Tekin I (2019) Monostatic antenna in-band full duplex radio: performance limits and characterization. IEEE Trans Veh Technol 68(5):4786–4799
25. Amjad MS, Nawaz H, Özsoy K, Gürbüz Ö, Tekin I (2018) A low-complexity full-duplex radio implementation with a single antenna. IEEE Trans Veh Technol 67(3):2206–2218
26. Korpi D, Choi YS, Huusari T, Anttila L, Talwar S, Valkama M (2015) Adaptive nonlinear digital self-interference cancellation for mobile inband full-duplex radio: algorithms and RF measurements. In: Proceedings of IEEE global communication conference (GLOBECOM), pp 1–7

27. Hassani SA, van Liempd B, Bourdoux A, Horlin F, Pollin S (2022) Joint in-band full-duplex communication and radar processing. IEEE Syst J 16(2):3391–3399
28. Lorenz KS, Goodman J, Mcbeth M, Mckeon MB, Parrett DJ (2022) A real-time wideband subband LMS algorithm for full-duplex communications. IEEE Access 10:32566–32573
29. Chung M, Sim MS, Kim J, Kim DK, Chae C (2015) Prototyping real-time full duplex radios. IEEE Commun Mag 53(9):56–63
30. Emara M, Rosson P, Roth K, Dassonville D (2017) A full duplex transceiver with reduced hardware complexity. In: Proceedings of IEEE global communication conference (GLOBECOM), pp 1–6
31. Korpi D, Heino M, Icheln C, Haneda K, Valkama M (2017) Compact inband full-duplex relays with beyond 100 dB self-interference suppression: enabling techniques and field measurements. IEEE Trans Antennas Propag 65(2):960–965
32. Anttila L, Lampu V, Hassani SA, Campo PP, Korpi D, Turunen M, Pollin S, Valkama M (2021) Full-duplexing with SDR devices: algorithms, FPGA implementation, and real-time results. IEEE Trans Wirel Commun 20(4):2205–2220
33. Yu B, Qian C, Lin P, Shao S, Pan W, Shen Y, Hu S, Su D, Sun C, Xiong Q, Lee J (2022) Full duplex communication with practical self-interference cancellation implementation. In: Proceedings of IEEE international conference on communication (ICC), pp 1100–1105
34. Yang J, Du C, Zhu Y, Ruan H, Zhang Z (2024) A parallel mechanism for fast digital SIC in full-duplex ISAC systems. In: Proceedings of IEEE international conference on signal information data processing (ICSIDP)
35. Hassani SA, Parashar K, Bourdoux A, van Liempd B, Pollin S (2019) Doppler radar with in-band full duplex radios. In: Proceedings of IEEE conference on computer communication (INFOCOM), pp 1945–1953
36. Hassani SA, Lampu V, Parashar K, Anttila L, Bourdoux A, van Liempd B, Valkama M, Horlin F, Pollin S (2021) In-band full-duplex radar-communication system. IEEE Syst J 15(1):1086–1097
37. Baquero Barneto C, Riihonen T, Turunen M, Anttila L, Fleischer M, Stadius K, Ryynänen J, Valkama M (2019) Full-duplex OFDM radar with LTE and 5G NR waveforms: challenges, solutions, and measurements. IEEE Trans Microw Theory Techn 67(10):4042–4054
38. Barneto CB, Liyanaarachchi SD, Heino M, Riihonen T, Valkama M (2021) Full duplex radio/radar technology: the enabler for advanced joint communication and sensing. IEEE Wirel Commun 28(1):82–88
39. Carusone AC, Johns DA (2002) Analog filter adaptation using a dithered linear search algorithm. In: Proceedings of IEEE international symposium circuits system (ISCAS), vol 4, pp IV-IV
40. Liyanaarachchi SD, Riihonen T, Barneto CB, Valkama M (2021) Optimized waveforms for 5G–6G communication with sensing: theory, simulations and experiments. IEEE Trans Wirel Commun 20(12):8301–8315

BY NC ND

Chapter 6
On-Chip RF Canceller

Abstract With the increasing demands of wireless communications, higher requirements of miniaturization and integration have been put forward for communication devices. Compared with the discrete-component-based implementations, on-chip RF cancellers would permit the design of commercially attractive compact FD ISAC devices [1], which is recognized as a key technology for large-scale applications by academia and industry. Accordingly, this chapter presents on-chip monolithic designs for RF SIC, including frequency-domain equalization (FDE) SIC architecture, time-domain equalization (TDE) SIC architecture, Hilbert-transform equalization (HTE) SIC architecture and MIMO SIC architecture, as depicted in Fig. 6.1. A detailed investigation is provided into key technologies, cancellation performance, as well as the advantages and disadvantages of each of these four architectures.

6.1 FDE Architecture

FDE technology segments the target bandwidth into multiple narrower sub-bands to mitigate the frequency-selective fading characteristics of the SI channel, as shown in Fig. 6.2a. Multiple RF bandpass filters (BPFs) with embedding variable attenuators and phase shifters are included in the canceller to channelize the desired signal bandwidth, thereby facilitating effective cancellation of the multi-carrier SI signal. For effective suppression of inter-channel interference between adjacent channels, the design of high-Q RF BPFs is critical as the high frequency selectivity and narrow passband of high-Q filters significantly enhance channel isolation.

In [2, 3], a 2-TAP FDE-based canceller was demonstrated, which achieved >20dB RF cancellation at 0.8∼1.4 GHz, over 20 MHz bandwidth with -20 dBm average input power and 64QAM modulated SI signal. N-path G_m-C RF BPFs were employed in their design. R_S and R_L represent the resistive loads at the TX and RX sides, respectively. C_C weakly couples the cancellation signal to the RX input for SIC. Given fixed R_S and R_L, the quality factor of an N-path filter can be reconfigured by adjusting the baseband capacitor C_B, enabling a quality factor of up to 81.2. Moreover, the intricate design tradeoff among RF cancellation performance, noise factor (NF) degradation and linearity was taken into consideration.

C. Du et al., *Full-Duplex Integrated Sensing and Communication Systems*,
https://doi.org/10.1007/978-981-92-0470-0_6

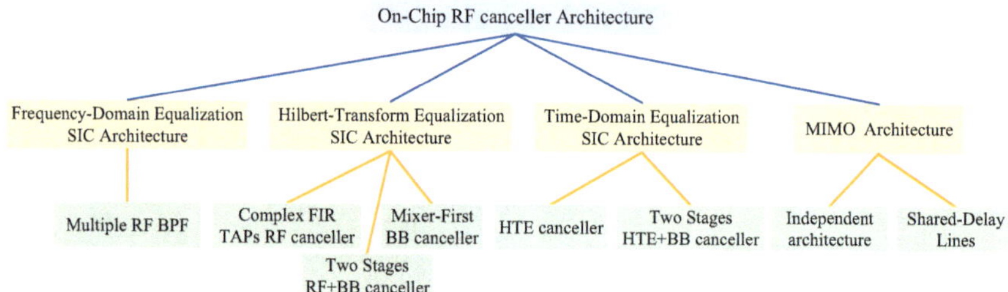

Fig. 6.1 On-chip RF canceller architecture

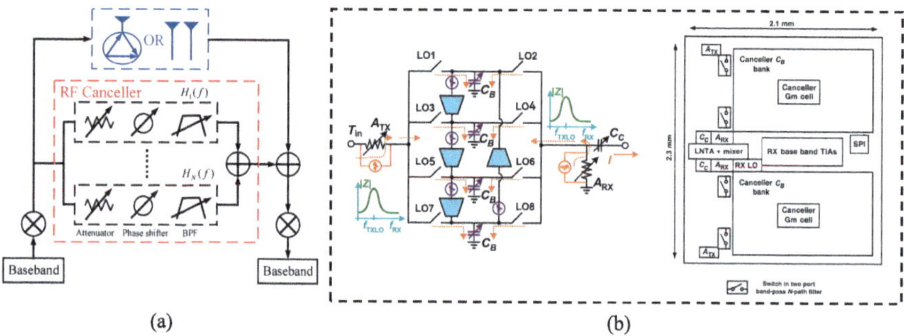

Fig. 6.2 **a** Architecture of FDE-based canceller; **b** Implemented BPFs based on Gm-C frequency shifted N-path filters and the chip design of the implemented canceler [2, 3].

Lessons Learned **:** The FDE architecture demonstrates efficacy in suppressing wideband SI signals characterized by frequency-selective fading. However, this architecture is confronted with the challenge of large chip area consumption caused by the capacitor banks, and the RF operating range is relatively narrow. Besides, the inter-channel interference cannot be completely eliminated, which imposes a fundamental limitation on the achievable RF cancellation performance.

6.2 TDE Architecture

TDE technology reconstructs the multipath SI using a complex finite impulse response (FIR) cancellation structure, where the channel response is simulated by TAPs with different delays and weights. Compared with FDE architecture, TDE architecture offers the advantage of avoiding inter-channel interference, but its implementation faces challenges in precise multipath group delay matching. TDE technology can be further classified into three categories based on architectural implementation and signal processing domain: complex FIR RF canceller, mixer-first baseband (BB) canceller and two stage (RF+BB) canceller.

6.2.1 Complex FIR RF Canceller

The architecture of complex FIR RF canceller is depicted in Fig. 6.3a. Each TAP incorporates a fixed delay, an attenuator and a phase shifter, which are employed to directly adjust the amplitude and phase of signals in RF domain, respectively.

In [4, 5], a single-TAP RF canceller and a pair of wideband polarization-based transceiver antennas were monolithically integrated on the same chip. A co-polarized auxiliary port was integrated into the receive antenna to establish an indirect coupling path between the transmitter output and receiver input, where a reconfigurable on-chip reflective termination was employed to reflect the coupled signal and cancel SI at the receiver input. With the combined implementation of antenna and RF cancellation, a total >70 dB cancellation was achieved over a 1 GHz bandwidth centered at 59GHz with +11dBm average transmitted power and BPSK modulated SI signal. Both attenuators and phase shifters are reflection-type, which is characterized by its simple structural configuration, but will introduce a higher level of nonlinear distortion.

Moreover, in [6], a polyphase filter (PPF) and tunable G_m stages were designed to implement a 360° phase shifter, which can significantly alleviate the loading effect on PA by increasing resistor and reducing capacitor, thereby further suppressing harmonic spurious output. The canceler employed a PPF to generate four approximately equal-amplitude signals with 0°, 90°, 180° and 270°phase shifts. Two signals with a 90° phase difference were selected and individually weighted using variable-gain transconductance stages, then combined to form a vector sum that matched the amplitude and phase of the leakage signal. This approach implemented a Cartesian rotator in the cancellation path to replicate the TX leakage. In their design, a single-TAP RF canceller can not only achieve 31 dB RF cancellation at 2.4G Hz over 4 MHz bandwidth with −15 dBm average input power and GFSK modulated SI signal, but also suppressed 3nd and 5th harmonic spurious output by 30 dB and 15 dB, respectively.

Although PPF-based design can mitigate nonlinear distortion, the incorporation of tunable G_m stages in [6] to achieve an attenuator may introduce non-flat impedance within the passband, leading to greatly mismatch at the receiver input, which limits the SIC bandwidth to only 4 MHz. To overcome this limitation, in [7], a passive capacitor digital to analog converter (C-DAC) attenuator was presented, which demonstrated dual improvements in both the in-band impedance flatness and the linearity of the canceller. The implemented single-TAP RF canceller provided >25dB cancellation across the 0.2∼3.8 GHz frequency range over 10MHz bandwidth with 64QAM modulated SI signal.

Given that only a single TAP is employed in the above canceller, the aforementioned complex FIR RF cancellers do not incorporate the delay lines in their design. In [8], two cancellers were proposed to place before and after the LNA. Each RF canceller consisted of a 5-TAP FIR filter and each TAP included an RC-CR allpass filter (APF) that provided a measured time delay of 50.9 ps. An EBD was also integrated on the chip, which can provide 39 dB antenna cancellation. When combined with antenna cancellation, the overall design provided a total 72.8/70.1/65.2 dB can-

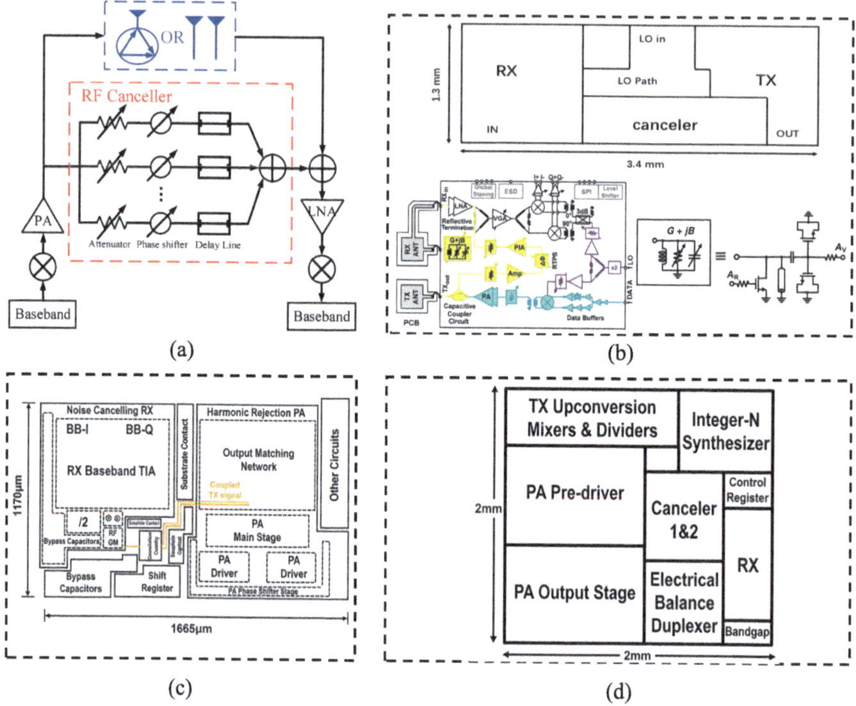

Fig. 6.3 **a** Architecture of complex FIR RF canceller. **b** Architecture of the FD 60 GHz TRX with reconfigurable polarization-based antenna and RF cancellation [4, 5]. **c** The chip design of the FD transceiver [6]. **d** Two RF feedforward FIR cancelers and FD transceiver die micrograph [8].

cellation on OFDM 64-QAM modulated signal at 1.6∼1.9 GHz, corresponding to integration bandwidths of 20/40/80 MHz.

Lessons Learned **:** In general, the multi-tap complex FIR RF canceller requires precise group delay adjustment to accurately match the characteristics of the multi-path SI signal. To achieve a cancellation level of up to 30∼40 dB for wideband SI, the delays provided by the cancellers should span across tens of nanoseconds due to the long delay spread of the SI response [9]. Nevertheless, the direct implementation of nanosecond-scale delay lines in RF domain is fundamentally constrained by the inverse relationship between time delay and frequency.

6.2.2 Mixer-First BB Canceller

The architecture of Mixer-First BB Canceller is depicted in Fig. 6.4a. Both SI signal and reference signal are down-converted to baseband, followed by the reconstruction of the SI signal through using a complex FIR BB filter. Compared with the RF Canceller, BB canceller offers greater flexibility in implementing delay adjustments at

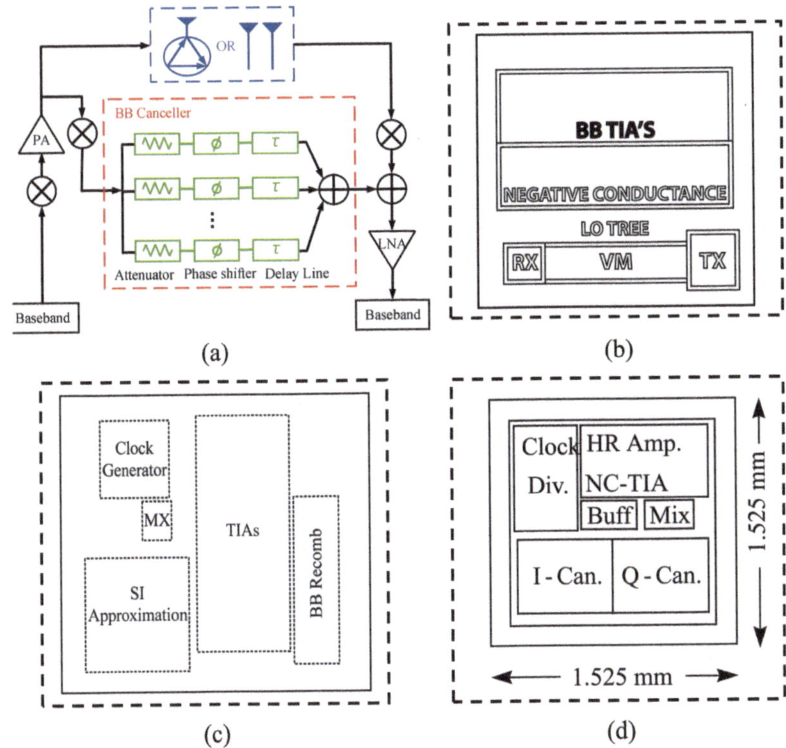

Fig. 6.4 **a** Architecture of Mixer-First BB Canceller. **b** The chip design of the implemented canceler [10–12]. **c** The chip design of the implemented canceller [15, 16]. **d** The chip design of the implemented canceller [14]

baseband. Moreover, it is crucial to adopt a mixer-first architecture by interchanging the positions of the LNA and mixer since the cancellation process has to take place before amplification due to the high power level of the SI signal. Meanwhile, the mixer-first architecture also demonstrates enhanced wideband linearity and superior NF performance.

In [10–12], a single-TAP mixer-first BB canceller was proposed. The design can achieve 27 dB RF cancellation across the frequency range of 0.15–3 GHz, over 16.25 MHz bandwidth. The NF was measured to be 6.3 dB in HD mode and 10.3–12.3 dB in FD mode. In their design, the phase shift, attenuation, and downmixing have been jointly integrated into a single component, namely a vector modulator (VM) downmixer. To strike a balance between the quantization error of attenuation and phase shift and practical on-chip implementation constraints, 31 slices of the VM were chosen, which can be conveniently segmented and controlled with 5 bits. Furthermore, in [13], their RF canceller chip was tested combined with a pair of dual-polarization antennas, which achieved 50 dB antenna cancellation and 12 dB RF cancellation at 2.46 GHz over 15 MHz bandwidth, using a 16QAM modulated

SI signal with an average transmitted power of +15 dBm and a peak group delay of 4 ns.

In [14], a passive continuous-mode charge-sharing VM downmixer was introduced to serve as the BB canceller. In the proposed design, the VM structure was implemented through the combination of two N-Path sampling mixers (SM) with variable attenuation, arranged in a quadrature configuration. Compared with prior works [10–13], their VM eliminated the requirement for a large number of slices and achieved a lower NF performance. The proposed single-TAP mixer-first BB canceller can achieve 27 dB RF cancellation at 0.1∼0.95 GHz over 15 MHz bandwidth with 1.2 ns peak group delay. The NF was measured to be 6.7∼7.9 dB in HD mode and 7.1∼8.8 dB in FD mode, where the degradation of NF introduced by transitioning from HD to FD operation was less than 1 dB.

However, the mixer-first BB cancellers discussed above exhibit an inherently high basic NF (> 7dB), which adversely degrades the performance of the receiver. In [15, 16], a single-TAP mixer-first BB canceller with low NF was proposed. Specifically, the NF was measured to be 3.3 dB in HD mode and 5.3 dB in FD mode. The canceller can achieve 35 dB RF cancellation at 0.5∼3.5 GHz, over 10 MHz bandwidth with $-25 \sim -65$ dBm average input power, 0.8∼2.5 group delay and 16QAM modulated SI signal. In their design, a Cartesian-based SI approximation incorporating a downmixer was utilized to adjust both the arbitrary amplitude and phase of the signals, effectively replacing the VM downmixer, which can reduce noise penalty. However, the implementation of in-phase quadrature (IQ) signal generation using resistor-inductor-capacitor (RLC) structure may lead to large chip area consumption.

Lessons Learned **:** Generally speaking, mixer-first BB cancellers with VM downmixer can mitigate SI signal with nanosecond-level group delay at the expense of increased chip area and high power consumption, which limits their suitability for compact and energy-efficient transceiver designs. Moreover, both RF and BB cancellers can only achieve approximately 30dB RF cancellation, which remains inadequate for effectively suppressing high-power SI signal.

6.2.3 Two Stage (RF+BB) Canceller

To overcome the limitations of RF cancellation, a two stage cancellation architecture can be employed in which RF and BB cancellers are cascaded before and after LNA respectively, as illustrated in Fig. 6.5a. In this architecture, RF canceller serves as the first stage of cancellation to provide small but finer group delays, while BB canceller operates as the second stage to provide large but coarse group delays.

In [17, 18], a two-stage SI cancellation architecture comprising both RF and BB cancellers was proposed, as illustrated in Fig. 6.5b. The design can achieve a total cancellation of 50.85 dB (32.76 dB in RF canceller and 18.09 dB in BB canceller) at 1.96 GHz, over 42 MHz bandwidth with -28.43 dBm average input power, 2 ns group delay, and 16QAM modulated SI signal. The proposed architecture was implemented

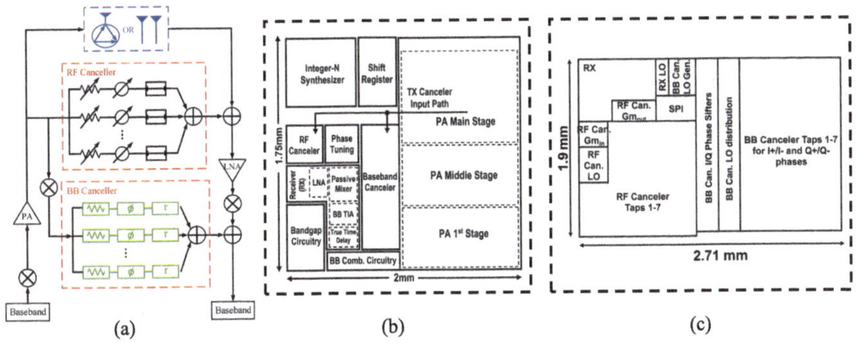

Fig. 6.5 **a** Architecture of two stage (RF+BB) Canceller. **b** System block of components on the 40-nm CMOS FD system based on RF and BB cancelers [17, 18]. **c** Detailed diagram of the FD receiver featuring time-interleaved switched-capacitor delay-based RF and BB cancelers [9, 19].

using: (1) a 5-TAP FIR filter in the RF canceller, which achieved 65 ps TAP-to-TAP delay by using RC-CR APF; (2) a 14-TAP low-frequency FIR filter using G_m-C based APF in the BB canceller, which offered a significantly higher Tap-to-Tap delay of 10 ns. The overall transceiver power consumption was 49 mW, excluding the integrated PA, and the fabricated prototype occupied a total chip die area of 3.5 mm^2.

In [9, 19], switched-capacitor delay circuits were proposed, which can offer much larger delays compared to RC-CR APF and lower power consumption compared to G_m-C based APF, thus demonstrating the feasibility of achieving large on-chip delays in RF and BB domains while maintaining low form factor and power consumption. In their design, as shown in Fig. 6.5c, 7-TAP FIR filters were employed in both RF canceller and BB canceller, with delay ranges of 0.2–1.1 ns and 10–75 ns, respectively, and gain control of 5 bits and 7 bits, respectively. Combined with a circulator to perform antenna cancellation, the canceller can achieve a total RF cancellation level of 52 dB, 49 dB, and 46 dB across bandwidths of 20 MHz, 30 MHz, and 40 MHz, respectively.

Lessons Learned : In general, the two stage (RF+BB) canceller can achieve higher performance of SI cancellation, typically $>$ 50dB, but this improvement is accompanied by increased power consumption and chip area overhead due to the added architectural components and complexity.

6.3 HTE Architecture

To overcome the challenge of realizing nanosecond delay lines in RF domain, prior works [20, 21] introduced an HTE architecture, in which RF delays can be effectively implemented by up-converting the BB delays, as illustrated in Fig. 6.6a. Since the delay needs to be applied only to the BB envelopes, it is more efficient to downconvert the RF signal to BB using mixers and a LPF to extract the desired components, then

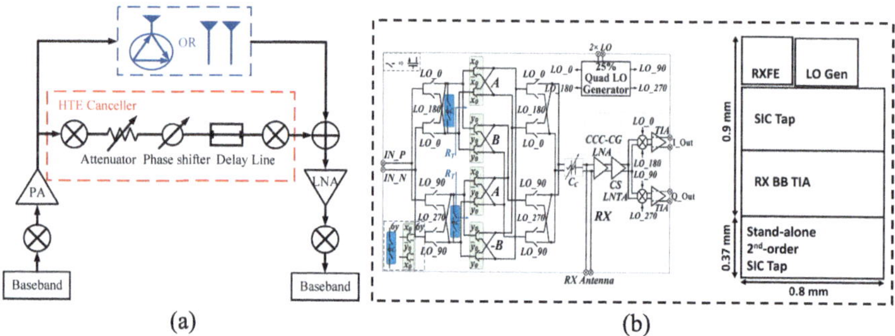

Fig. 6.6 **a** Architecture of HTE canceller. **b** Simplified schematic of the proposed HTE based SIC filter and the chip design of the implemented canceler [20, 21]

apply the delay in the BB domain, and subsequently upconvert the signal back to RF. In the proposed design, the RF canceller contained a single-TAP with 1.4 ns group delay by using two LPFs, which achieved a total 73 dB cancellation (50 dB in antenna domain and 23 dB in RF domain) at 900 MHz, over 80 MHz bandwidth with +11 dBm average transmitted power and 16QAM modulated SI signal.

Furthermore, a two stage canceller architecture was proposed in [22, 23], where a single-TAP HTE canceller was placed before the LNA and a BB canceller was deployed after it, as illustrated in Fig. 6.7. In the proposed FD receiver design, a capacitor stacking (CS)-based second-order delay cell was employed to break the trade-offs between delay, loss, circuit area, and noise in HTE canceller. A prototype of the proposed design was implemented in 65 nm Complementary Metal Oxide Semiconductor (CMOS) process. Overall, the canceller achieved a group delay ranging from 2∼8.8 ns and 9∼15 ns in HTE and BB domain, respectively. Their canceller achieved a total 33 dB cancellation (27 dB in HTE canceller and 6 dB in BB canceller) at 900 MHz, over 20 MHz bandwidth with 64QAM modulated SI signal. The receiver power handling capability was improved by 11.5 dB, while the active chip area was limited to only 0.4 mm^2.

Lessons Learned : Compared to the complex FIR RF canceller, HTE canceller downconverts the RF SI to baseband, where it becomes more feasible to implement nanosecond delay lines by operating at low-frequency. However, in practical transceiver implementation, the HTE canceller requires an additional buffer circuit (such as a source follower) at the input to mitigate its impact on the PA efficiency, which in turn limits the linearity of the canceller. Moreover, HTE cancellers generally incur higher power consumption and occupy larger chip area compared to TDE cancellers, making them less favorable for integration in compact or energy-constrained systems.

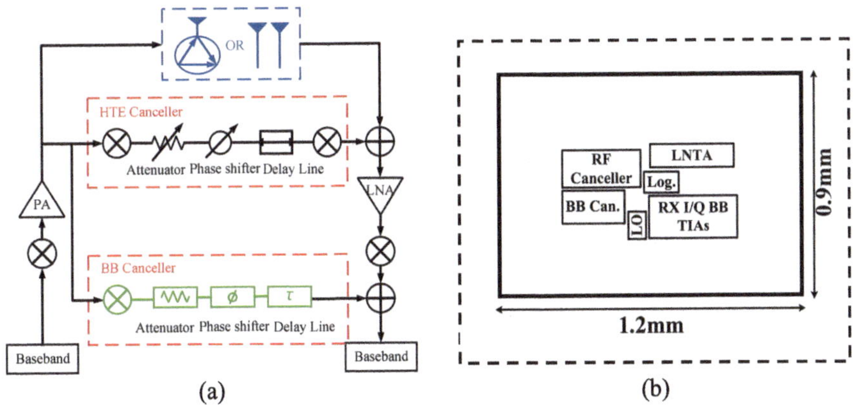

Fig. 6.7 **a** Architecture of two stage (HTE+BB) canceller. **b** The chip diagram of the implemented canceler [22, 23]

6.4 MIMO Architecture

The application of FD technology to MIMO systems introduces new challenges for transceiver design, as it must simultaneously suppress SI from each TX to its own RX, and cross-talk interference (XI) between the antennas. Figure 6.8a illustrates the architecture of 2×2 FD MIMO transceiver, where two cancellers are required for SIC and additional two cancellers are required for XI cancellation (XIC). In general, the cancellation complexity of a $N \times N$ FD-MIMO transceiver scales as $O(N^2)$, where N SI cancellers and $N \times (N-1)$ XI cancellers are required. Although each canceller can be integrated separately, it will inevitably lead to an increase in the overall form factor. Therefore, the ultimate goal of RF MIMO transceiver design is to achieve monolithic integration of RF cancellation functionality on a single chip.

In [24, 25], a 2×2 MIMO incorporating four mixer-first BB cancellers was demonstrated, as shown in Fig. 6.8b. The design achieved a total 24 dB cancellation

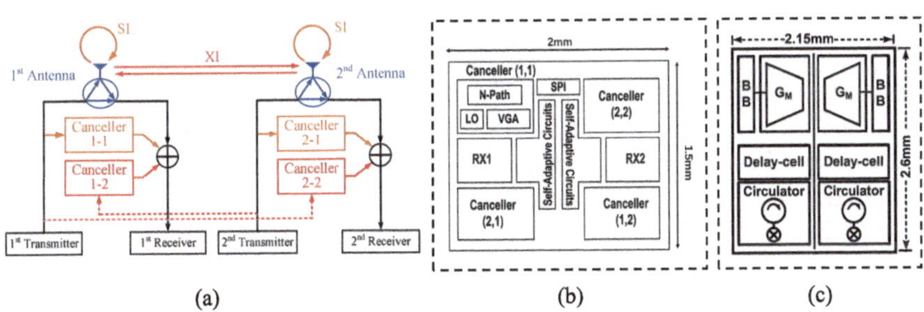

Fig. 6.8 **a** MIMO architecture. **b** The chip diagram of the implemented canceler [24, 25]. **c** The chip diagram of the implemented canceler [26, 27]

(6 dB for XIC and 18 dB for SIC), at 0.5∼2.5 GHz over 20 MHz bandwidth with maximum -5dBm average input power and QPSK modulated SI signal. In their design, complex LMS adaptation loops were also integrated on the chip to adaptively tune the weights of complex FIR filter. To reduce adaptation overhead, the LMS adaptive circuitry was co-designed with and partially embedded in a wideband RF/analog SIC and a gain-boosted mixer-first receiver (RX). This approach minimized the number of canceller taps and enabled functional integration within a single circuit block.

However, as the number of cancellers escalates quadratically with the increase in the number of antennas, the complexity of G_m-C cells of delay lines also increases quadratically, which will lead to prohibitive hardware costs. To overcome this fundamental limitation, prior works [26, 27] proposed the use of shared-delay lines to reduce the complexity from $O(N^2)$ to $O(N)$. G_m-cell-based vector modulators tapped from the discrete delay steps of each TX element and injected into all RX BB paths to approximate the SI channel. Since each TX path employed a single delay line with its taps shared among all RX paths, the number of required delay lines scaled linearly with the number of MIMO elements rather than quadratically. They proposed a 2×2 MIMO system equipped with four two stage (RF+BB) cancellers, which can achieve a total 53 dB cancellation (8dB for XIC and 45dB for SIC) at 2.2 GHz, over 20 MHz bandwidth.
Lessons Learned **:** Generally speaking, the $O(N^2)$ scaling of complexity, chip area and power consumption, etc., poses significant challenges that hinder the practical deployment and scalability of FD-ISAC MIMO systems. As the number of antennas increases, the quadratic growth in hardware and computational requirements imposes substantial burdens on circuit integration, thermal management, and energy efficiency. Moreover, it becomes increasingly challenging to jointly tune the weights for both SI and XI cancellers, primarily due to the compounded effects of hardware impairments and multipath propagation.

In summary, Table 6.1 describes the performance of recently published on-chip RF cancellers.

6.5 Conclusion

In conclusion, this chapter has provided a comprehensive overview of circuit approaches for RF SIC in FD systems. Various aspects of hardware design and implementation focusing on on-chip RF cancellers are discussed, which can be classified into four architectural types, including FDE, TDE, HTE and MIMO architecture. Moreover, this chapter also offers a comparative analysis of the performance of the state-of-the-art RF cancellers based on these four architectures, highlighting their respective advantages and trade-offs in terms of cancellation performance, bandwidth, NF degradation and chip area. Looking ahead, a critical direction for future research lies in the development of highly reconfigurable, low-power, and area-efficient RF cancellation circuits that are compatible with emerging FD applications such as ISAC.

Table 6.1 Performance summary and comparison of on-chip RF cancellers

Work	Year	Architecture	Frequency	Bandwidth	Transmit Power	Antenna cancellation	RF cancellation	Rx N/ degradation in FD mode	Technology /Area
[2, 3]	2015	FDE	0.8~1.4 GHz	27 MHz	+15 dBm	35 dB	20 dB	4.8 dB/1.3 dB	65 nm/ 4.8 mm^2
[4, 5]	2016	TDE Complex FIR	59 GHz	1 GHz	+11 dBm	>70 dB		4 dB/–	40 nm/12.5 mm^2
[8]	2018	TDE Complex FIR	1.6~1.9 GHz	20/40/80 MHz	–	72.8/70.1/65.2 dB		8.1 dB/1.6 dB	40 nm/4 mm^2
[6]	2017	TDE Complex FIR	2.4 GHz	4 MHz	+20 dBm	35 dB	31 dB	4.6 dB/0.6 dB	40 nm/0.15 mm^2
[7]	2022	TDE Complex FIR	0.2~3.8 GHz	10 MHz	–	30 dB	30 dB	3.2~5.9 dB/ 0.85~1.5 dB	65 nm/0.55 mm^2
[10–13]	2015	TDE Mixer-First BB	0.15~ 3GHz	16.25 MHz	+0.2 dBm	29 dB	27 dB	6.3 dB/ 4~6dB	65 nm/2 mm^2
[14]	2019	TDE Mixer-First BB	0.1~0.95 GHz	15 MHz	–	–	27 dB	7.1~8.8 dB/ 0.4~0.9 dB	130 nm/1.4 mm^2
[15, 16]	2020	TDE Mixer-First BB	0.5~3.5 GHz	10 MHz	–	–	35 dB	3.3 dB/2 dB	65 nm/1.5 mm^2
[17, 18]	2018	TDE Two Stage	1.7~2.2 GHz	42 MHz	+15 dBm	43.43 dB	50.85 dB	-/1.55 dB	40 nm/3.5 mm^2
[9, 19]	2021	TDE Two Stage	0.1~1 GHz	20/30/40 MHz	+9 dBm	22/22/23 dB	30/27/23 dB	5.3 dB/1.9 dB	65 nm/5.15 mm^2
[20, 21]	2019	HTE	900 MHz	80 MHz	+6 dBm	50 dB	23 dB	9.6 dB/1.4 dB	130 nm/0.72 mm^2
[22, 23]	2023	HTE Two Stage	0.5~4 GHz	20 MHz	+12.5 dBm	23~26 dB	32~38 dB	3.7~5.7 dB/ 1.3 dB	65 nm/0.4 mm^2
[24, 25]	2020	MIMO	0.5~2.5 GHz	20 MHz	–	24 dB	18 dB	3.1~6 dB/ 1.3 dB	65 nm/3 mm^2
[26, 27]	2019	MIMO	2.2 GHz	20 MHz	–	53 dB		9.5 dB/2.1 dB	65 nm/5.6 mm^2

References

1. Debaillie B, van den Broek DJ, Lavín C, van Liempd B, Klumperink EAM, Palacios C, Craninckx J, Nauta B, Pärssinen A (2014) Analog/RF solutions enabling compact full-duplex radios. IEEE J Sel Areas Commun 32(9):1662–1673
2. Zhou J, Chuang T-H, Dinc T, Krishnaswamy H (2015) Integrated wideband self-interference cancellation in the RF domain for FDD and full-duplex wireless. IEEE J Solid-State Circuits 50(12):3015–3031
3. Zhou J, Chuang T-H, Dinc T, Krishnaswamy H (2015) Receiver with >20 MHz bandwidth self-interference cancellation suitable for FDD, co-existence and full-duplex applications. In: Proceedings of IEEE international solid-state circuits conference (ISSCC), pp 1–3
4. Dinc T, Chakrabarti A, Krishnaswamy H (2015) A 60 GHz same-channel full-duplex CMOS transceiver and link based on reconfigurable polarization-based antenna cancellation. In: Proceedings IEEE radio frequency integrated circuits symposium (RFIC), pp 31–34
5. Dinc T, Chakrabarti A, Krishnaswamy H (2016) A 60 GHz CMOS full-duplex transceiver and link with polarization-based antenna and RF cancellation. IEEE J Solid-State Circuits 51(5):1125–1140
6. Zhang T, Chen Y, Huang C, Rudell JC (2017) A low-noise reconfigurable full-duplex front-end with self-interference cancellation and harmonic-rejection power amplifier for low power radio applications. In: Proceedings European solid-state circuits conference (ESSCIRC), pp 336–339
7. Wang C, Li W, Xu H (2021) A 0.2-3.8-GHz full-duplex receiver with more than 25 dB self-interference cancellation using a C-DAC-based vector canceller. IEEE Access 9:148574–148589
8. Katanbaf CK-D, Zhang M, Su T, Rudell C, JC (2018) A broadband and deep-TX self-interference cancellation technique for full-duplex and frequency-domain-duplex transceiver applications. In: IEEE international solid-state circuits conference (ISSCC), pp 170–172
9. Nagulu A, Gaonkar A, Ahasan S, Garikapati S, Chen T, Zussman G, Krishnaswamy H (2021) A full-duplex receiver with true-time-delay cancelers based on switched-capacitor-networks operating beyond the delay-bandwidth limit. IEEE J Solid-State Circuits 56(5):1398–1411
10. van den Broek D-J, Klumperink EAM, Nauta B (2015) 19.2 A self-interference-cancelling receiver for in-band full-duplex wireless with low distortion under cancellation of strong TX leakage. In: 2015 IEEE international solid-state circuits conference (ISSCC) digital technology paper, pp 1–3
11. van den Broek D-J, Klumperink EAM, Nauta B (2015) A self-interference cancelling front-end for in-band full-duplex wireless and its phase noise performance. In: 2015 IEEE radio frequency integrated circuits symposium (RFIC), pp 75–78
12. van den Broek D-J, Klumperink EAM, Nauta B (2015) An in-band full-duplex radio receiver with a passive vector modulator downmixer for self-interference cancellation. IEEE J Solid-State Circuits 50(12):3003–3014
13. Debaillie B, van den Broek DJ, Lavín C, van Liempd B, Klumperink EAM, Palacios C, Craninckx J, Nauta B (2015) RF self-interference reduction techniques for compact full duplex radios. In: 2015 IEEE 81st vehicular technology conference (VTC Spring), pp 1–6
14. Sharma PK, Nallam N (2019) A 0.1-0.95 GHz full-duplex receiver with 1 dB NF degradation using a passive continuous-mode charge-sharing vector modulator. IEEE Trans Microw Theory Technol 67(7): 3042–3052
15. Ershadi A, Entesari K (2020) A 0.5-to-3.5-GHz full-duplex mixer-first receiver with Cartesian synthesized self-interference suppression interface in 65-nm CMOS. IEEE Trans Microw Theory Technol 68(6):1995–2010
16. Ershadi A, Entesari K (2019) A 0.5-to-3.5 GHz self-interference-canceling receiver for in-band full-duplex wireless. In: 2019 IEEE radio frequency integrated circuits symposium (RFIC), pp 151–154

17. Zhang T, Su C, Najafi A, Rudell JC (2018) Wideband dual-injection path self-interference cancellation architecture for full-duplex transceivers. IEEE J Solid-State Circuits 53(6):1563–1576
18. Zhang T, Najafi A, Su C, Rudell JC (2017) 18.1 A 1.7-to-2.2GHz full-duplex transceiver system with $>50dB$ self-interference cancellation over 42MHz bandwidth. In: 2017 IEEE international solid-state circuits conference (ISSCC), pp 314–315
19. Nagulu A, Gaonkar A, Ahasan S, Chen T, Zussman G, Krishnaswamy H (2020) A full-duplex receiver leveraging multiphase switched-capacitor-delay based multi-domain FIR filter cancelers. In: 2020 IEEE radio frequency integrated circuits symposium (RFIC), pp 43–46
20. El Sayed A, Mishra AK, Ahmed AH, Shirazi AHM, Woo S-P, Choi Y-S, Mirabbasi S, Shekhar S (2019) A Hilbert transform equalizer enabling 80 MHz RF self-interference cancellation for full-duplex receivers. IEEE Trans Circuits Syst I Regul Pap 66:1153–1165
21. El Sayed A, Ahmed A, Mishra AK, Shirazi AHM, Woo S-P, Choi Y-S, Mirabbasi S, Shekhar S (2017) A full-duplex receiver with 80 MHz bandwidth self-interference cancellation circuit using baseband Hilbert transform equalization. In: 2017 IEEE radio frequency integrated circuits symposium (RFIC), pp 360–363
22. Wang C, Li W, Chen F, Zuo W, Pu Y, Xu H (2022) A 0.5-4 GHz full-duplex receiver with multi-domain self-interference cancellation using capacitor stacking based second-order delay cells in RF canceller. In: 2022 IEEE radio frequency integrated circuits symposium (RFIC), pp 259–262
23. Wang C, Ma X, Li W, Chen F, Zuo W, Pu Y, Zhou J, Xu H (2024) A wideband full-duplex receiver with multi-domain self-interference cancellation based on capacitor stacking delay and delay compensation in cancellers. IEEE J Solid-State Circuits 59:1697–1708
24. Cao Y, Zhou J (2019) A CMOS 0.5-2.5 GHz full-duplex MIMO receiver with self-adaptive and power-scalable RF/analog wideband interference cancellation. In: IEEE radio frequency integrated circuits symposium (RFIC), pp 147–150
25. Cao Y, Zhou J (2020) Integrated self-adaptive and power-scalable wideband interference cancellation for full-duplex MIMO wireless. IEEE J Solid-State Circuits 55:2984–2996
26. Baraani Dastjerdi M, Jain S, Reiskarimian N, Natarajan A, Krishnaswamy H (2019) Analysis and design of a full-duplex two-element MIMO circulator-receiver with high TX power handling exploiting MIMO RF and shared-delay baseband self-interference cancellation. IEEE J Solid-State Circuits 54:3525–3540
27. Dastjerdi MB, Jain S, Reiskarimian N, Natarajan A, Krishnaswamy H (2019) 28.6 Full-duplex 2×2 MIMO circulator-receiver with high TX power handling exploiting MIMO RF and shared-delay baseband self-interference cancellation. In: 2019 IEEE international solid-state circuits conference (ISSCC), pp 448–450

Part III
Current Challenges and Emerging Frontiers in FD-ISAC

Chapter 7
Challenge for FD-ISAC

Abstract This chapter, based on previous discussions, provides an in-depth analysis of the main challenges and development trends of FD-ISAC technology within current research. In terms of model design, the discussion addresses deficiencies in performance metrics, performance boundaries, precise modeling of self-interference, joint waveform design, and integrated channel models. For signal processing, this chapter also explores issues regarding echo signal recognition, the trade-off between algorithm accuracy and complexity, and resource allocation. Concerning FD-ISAC implementation and chip integration, this chapter highlights current challenges such as frequency and bandwidth limitations, as well as complexity and power consumption issues in the chip process. Research on FD-ISAC systems is still in its early stages. Although many concepts have been proposed and analyzed through simulations, and some preliminary experiments have been conducted, experimental research remains basic. Current studies face several challenges: model design, signal processing, realization and on-chip implementation.

7.1 FD-ISAC Model Design

7.1.1 Performance Metrics

In conventional communication systems, link performance has long been assessed through metrics such as achievable data rate, BER, SNR, and spectral efficiency, each illuminating a different aspect of throughput and reliability. Similarly, sensing platforms are judged by parameters like range and angular resolution, detection probability, false alarm rate, and estimation accuracy, which together describe a system's capacity to detect, localize, and characterize targets [1]. Historically, these two domains have evolved in parallel: communication engineers have pursued ever-higher throughput and lower BER via advanced coding schemes and multi-antenna techniques, while radar and lidar researchers have refined pulse designs to enhance spatial resolution and detection robustness. In resource-rich environments—where ample spectrum, power, and hardware are available—this isolationist approach has proven sufficient.

C. Du et al., *Full-Duplex Integrated Sensing and Communication Systems*,
https://doi.org/10.1007/978-981-92-0470-0_7

Nonetheless, as spectrum becomes increasingly congested and hardware platforms are called upon to perform both communications and sensing, independently tuning each metric can lead to unintended trade-offs [2–4]. For example, adopting higher-order modulation or aggressive beamforming to boost data rates may shorten dwell time on targets or introduce waveform ambiguities that impair sensing. Conversely, deploying radar-optimized pulses for fine resolution can monopolize bandwidth and degrade BER when carrying information. These intertwined effects reveal the need for a unified evaluation framework in FD-ISAC systems.

Instead of treating throughput and error performance on one side and detection probability and resolution on the other as separate objectives, future work should establish joint metrics that map out the trade-off frontier between communication capacity and sensing fidelity. Potential directions include extending classical information-theoretic capacity bounds to incorporate estimation error, deriving Pareto-optimal waveform designs that explicitly balance bits per second against estimation accuracy, and proposing composite figures of merit—such as "sensing throughput," which integrates data rate with detection confidence. By benchmarking FD-ISAC architectures against such holistic criteria, designers can allocate spectrum, power, and hardware resources in a way that simultaneously ensures robust data links and precise situational awareness.

7.1.2 Performance Limits for FD-ISAC

Full-duplex ISAC platforms aim to merge high-speed data transfer with radar-grade target detection, but field deployments routinely fall short of their theoretical ceilings. This shortfall stems from the very resource constraints that practical systems must endure—finite transmit power and bandwidth, hardware imperfections such as phase noise and amplifier nonlinearity, and tight size, weight, and cost budgets. Fundamental-limit studies map out the Pareto frontier linking sensing fidelity (e.g., range-estimation error, angular resolution, detection probability) to communication performance (e.g., achievable rate, BER, spectral efficiency) under shared resource caps. By casting power, spectrum, and hardware complexity as jointly scarce, these analyses yield closed-form bounds or convex-optimization formulations that prescribe the optimal joint design of waveform parameters, beam patterns, and time-frequency allocations.

Early investigations—such as those in [5, 6] for half-duplex downlink ISAC, and the stochastic-model treatment in [7]—have illuminated key trade-offs but typically overlook the self-interference inherent in full-duplex operation, assume perfect transmit-receive isolation, or simplify sensing to a single static target. To narrow the chasm between theory and practice, future work must embed realistic self-interference models, account for hardware impairments, and embrace multi-target dynamics within unified optimization frameworks. Such efforts will sharpen the Pareto boundary for FD-ISAC and inspire adaptive strategies—waveform shaping, power splitting, beam scheduling—that can track these bounds in real time. Ultimately, a holistic FD-ISAC performance model will underpin resource-allocation schemes capable of delivering near-optimal communication throughput and sensing precision, even under stringent hardware and spectral constraints.

7.1.3 Accurate Modeling of Self-Interference Signals

Effectively deploying FD-ISAC hinges on our capacity to accurately characterize and then cancel the powerful SI signal that leaks from transmitter to receiver. The challenge is especially acute for ultra-wideband (UWB) operation in rich multipath environments, where reflections, diffractions, and scatterers generate numerous delayed echoes of the transmitted waveform. Studies have shown these reflections can stretch out over hundreds of nanoseconds, vastly complicating SI cancellation[8]. First, modeling such a long, time-dispersed channel demands estimators of immense complexity and computational heft. Second, reconstructing and subtracting the composite interference in real time pushes existing hardware and algorithms to their limits.

Moreover, the very breadth of the UWB spectrum magnifies these difficulties: as delay spread grows, the instantaneous bandwidth of the summed SI signal widens, intensifying frequency-selective fading across the band. In practice, this means different frequency slices encounter distinct gains and phase shifts, thwarting any single-filter solution and forcing engineers to either piece together multiple narrowband cancellers or pursue more intricate, wideband designs. Together, the need for long-horizon channel estimation and the severe frequency selectivity inherent to high-delay UWB channels create a daunting barrier to achieving the SI suppression levels required for practical full-duplex performance. Breaking through will require advances not only in adaptive filtering and channel-modeling techniques but also in hardware architectures engineered to sustain these extreme signal dynamics.

7.1.4 Joint Waveform Design Optimization

Designing a unified waveform that simultaneously enables high-throughput data transmission, precise environmental sensing, and robust SIC remains a formidable challenge in FD-ISAC research. Although early work such as that in [9] hints at a beneficial link between stronger interference suppression and enhanced radar performance, these studies typically overlook the broader impact on communications metrics like throughput, error rate, and latency under joint operation. In practice, the attributes that maximize SI nulling—for example, aggressive spectral shaping or extreme beam-steering—could unintentionally undermine range resolution or reduce effective data capacity.

7.2 FD-ISAC Signal Processing

7.2.1 Tradeoff Between Target RFSIC and DSIC

The interplay between RFSIC and digital DSIC represents a fundamental bottleneck in FD transceiver design. On the one hand, RFSIC must be sufficiently aggressive to drive the residual SI power down into the linear operating range of the ADC, preventing saturation and excessive quantization noise in the digital domain. On the other hand, several studies have demonstrated that pushing analog cancellation too far can inadvertently suppress the dominant linear SI components while leaving—and even relatively magnifying—nonlinear distortions, mixer phase noise, and other hardware impairments. These unwanted residuals then undermine the performance of DSIC algorithms, which rely on an accurate linear-SI model for subtraction [8, 10]. To address this dilemma, recent works have proposed adaptive hybrid architectures: frequency-selective analog filters that shape the SI spectrum before digitization, machine-learning-driven tuning of cancellation parameters in real time, and decision-directed loopback schemes that dynamically allocate cancellation effort between the RF and digital domains based on instantaneous channel and hardware conditions.

Balancing RFSIC and DSIC for optimal overall performance remains a challenging task. Future transceiver designs must calibrate the depth of analog cancellation to sufficiently reduce SI power without exacerbating nonlinear artifacts that impede digital subtraction. Promising directions include joint analog-digital optimization algorithms that treat cancellation as a single unified problem, real-time monitoring of residual error statistics to adjust cancellation weights on the fly, and closed-loop control schemes that automatically rebalance RFSIC and DSIC as operating conditions evolve. By embracing holistic, system-level approaches—ones that account for hardware impairments, channel variability, and algorithmic complexity in concert—robust high-throughput full-duplex operation can be realized.

7.2.2 Echo Signal Identification

In realistic FD-ISAC deployments, echo-signal identification must contend with far more complexity than simple target reflections. In cluttered environments, transmitted waves interact with myriad objects, producing not only direct echoes but also refracted and scattered components that can masquerade as legitimate returns. As noted in [11, 12], these unintended refracted waves—originating from non-target surfaces—arrive at the receiver alongside genuine echoes, effectively acting as dynamic clutter. Meanwhile, many ISAC platforms rely on waveforms crafted for high data throughput rather than ideal autocorrelation characteristics. Unlike traditional radar pulses, which exhibit sharp correlation peaks and minimal sidelobes for precise range and velocity estimation, data-embedded ISAC signals include pseudo-random sequences that degrade these correlation properties [13]. The result is a significantly reduced ability to discriminate weak target echoes from both environmental clutter and residual self-interference when using conventional matched-filter processing.

This tangled signal landscape—comprising target reflections, refracted interference, waveform-induced ambiguity, and possible residual SI in FD operation demands novel processing strategies. Future research should therefore focus on developing advanced algorithms capable of isolating authentic target echoes within this complex milieu. Potential directions include hybrid correlation/detection schemes that compensate for suboptimal waveform auto-correlations, adaptive clutter-suppression filters informed by environmental models, and machine-learning approaches trained to recognize and reject non-target signal features. By addressing both environmental and system-originated interference in an integrated manner, these techniques will be crucial to achieving reliable, high-resolution sensing in next-generation FD-ISAC systems.

7.2.3 Tradeoff Between Accuracy and Complexity

When designing signal processing algorithms for ISAC, it is essential to find a reasonable balance between complexity and accuracy. In ISAC systems, the design of signal processing algorithms is crucial as they directly impact the overall system performance and the feasibility of practical applications. High-precision signal processing algorithms can provide more accurate communication and sensing functions, such as higher data transmission rates, lower bit error rates, and increased sensing accuracy. However, these high-precision algorithms often come with higher computational complexity and resource consumption, which may increase hardware requirements, power consumption, and response times, thereby limiting their deployment and widespread adoption in practical applications. On the other hand, low-complexity signal processing algorithms have clear advantages in terms of computational resources, power consumption, and response time, making them easier to implement in resource-constrained environments. However, this simplification often comes at the cost of sacrificing some degree of accuracy, which may result in decreased communication data rates, increased bit error rates, or reduced sensing accuracy [14]. Therefore, the challenge lies in finding an optimization solution that allows the system to maintain sufficient accuracy while minimizing complexity as much as possible.

7.3 FD-ISAC Realization and On-Chip Implementation

7.3.1 Frequency and Bandwidth Limitations

A salient limitation of current FD-ISAC prototypes is their reliance on relatively narrow signal bandwidths, which constrains sensing performance. Fine range resolution—and therefore precise target localization and discrimination—depends inversely on the transmitted bandwidth. Yet, FD hardware, which underpins many single-antenna ISAC schemes, struggles to deliver the level of SIC and RF component performance required for truly wideband operation. Although early demonstrations have confirmed the feasibility of in-band FD radios under controlled setting, these proof-of-concept systems do not yet confront the full complexity of wideband self-interference mitigation.

As a result, a clear gap persists between the broad spectral apertures needed for high-fidelity ISAC sensing and the present capabilities of FD-ISAC platforms. Bridging this divide demands concerted advances in both SIC algorithms and wideband RF design. Future work must rigorously characterize and improve SIC over expansive frequency ranges, optimize broadband RF front ends, and validate long-term stability and robustness in real-world environments. Only through such efforts can FD-ISAC systems fully exploit wideband resources to achieve their sensing and communication potential.

7.3.2 Embedded Circuits for Adaptive Weights Tuning

Employing adaptive algorithms is essential for maintaining optimal RFSIC performance in environments where the self-interference channel varies over time. Conventional implementations rely on digital-domain adaptation: the RF signals are down-converted to baseband, digitized, and then used to update the coefficients of multi-tap FIR filters (often via LMS or RLS routines) that drive digital cancellation or predistortion stages. However, realizing wideband cancellation in this fashion typically demands multiple parallel down-converter chains—each comprising mixers, filters, and ADCs—which can draw on the order of 1.5 W per chain, as noted in [15]. Such power requirements severely limit applicability in battery-powered devices like mobile handsets or sensor nodes.

An emerging alternative embeds the adaptation loop directly in the RF (or IF) domain, tuning complex filter weights through analog components—vector modulators, variable gain amplifiers, or phase shifters—rather than full baseband conversion, as explored in [16]. By eliminating multiple down-conversion paths and their associated ADCs, this approach promises dramatic reductions in both circuit complexity and power consumption, making advanced RFSIC viable for power-sensitive platforms.

Nonetheless, practical realization of embedded RF-domain adaptation presents its own challenges. Designing analog adaptive loops that remain stable, accurate, and robust across process, voltage, and temperature variations is nontrivial. Further research is therefore required to refine circuit topologies, improve loop stability, and validate long-term performance under real-world dynamics. Only then can RF-integrated adaptive weight tuning overcome the power and complexity limitations of traditional baseband methods and enable truly portable RFSIC solutions.

7.4 Conclusion

While the aforementioned studies highlight the cutting-edge advancements and broad prospects of FD-ISAC technology, this chapter focuses on the potential challenges it currently faces. Specifically, limitations exist in critical aspects such as channel modeling accuracy, signal processing efficiency, and hardware implementation costs. These technical bottlenecks, to varying degrees, affect the large-scale deployment and engineering applications of FD-ISAC systems. It is important to emphasize that acknowledging these technical challenges does not negate the research value or development potential of FD-ISAC. On the contrary, the research direction established through the systematic problem analysis in this chapter provides a clear focus for the "Future Directions for FD-ISAC" discussed in subsequent chapters. Breaking through these unresolved problems will not only promote the improvement of the FD-ISAC technology system but also lay an important foundation for the leapfrog development of future ISAC technologies.

References

1. Liu F, Cui Y, Masouros C, Xu J, Han TX, Eldar YC, Buzzi S (2022) Integrated sensing and communications: toward dual-functional wireless networks for 6g and beyond. IEEE J Sel Areas Commun 40(6):1728–1767
2. Cao N, Chen Y, Gu X, Feng W (2020) Joint radar-communication waveform designs using signals from multiplexed users. IEEE Trans Commun 68(8):5216–5227
3. Xu Z, Liu F, Petropulu A (2022) Cramér-Rao bound and antenna selection optimization for dual radar-communication design. In: Proceedings of ICASSP 2022 - 2022 IEEE international conference on acoustic speech signal processing (ICASSP), pp 5168–5172
4. Hassanien A, Amin M G, Zhang Y D, Ahmad F (2015) Dual-function radar-communications using phase-rotational invariance. In: Proceedings of 2015 23rd European signal processing conference (EUSIPCO), pp 1346–1350
5. Ouyang C, Liu Y, Yang H (2023) MIMO-ISAC: performance analysis and rate region characterization. IEEE Wirel Commun Lett 12(4):669–673
6. Liu A, Huang Z, Li M, Wan Y, Li W, Han TX, Liu C, Du R, Tan DKP, Lu J, Shen Y, Colone F, Chetty K (2022) A survey on fundamental limits of integrated sensing and communication. IEEE Commun Surv Tuts 24(2):994–1034

7. Dong F, Liu F, Lu S, Xiong Y, Yuan W, Cui Y (2024) Fundamental Limits of Communication-Assisted Sensing in ISAC Systems. In: Proceedings of 2024 IEEE international symposium information theory (ISIT), pp 2586–2591
8. Zhang Z, Long K, Vasilakos AV, Hanzo L (2016) Full-duplex wireless communications: challenges, solutions, and future research directions. IEEE Proc 104(7):1369–1409
9. He Z, Xu W, Shen H, Ng DWK, Eldar YC, You X (2023) Full-duplex communication for ISAC: joint beamforming and power optimization. IEEE J Sel Areas Commun 41(9):2920–2936
10. Mohammadi M, Mobini Z, Galappaththige D, Tellambura C (2023) A comprehensive survey on full-duplex communication: current solutions, future trends, and open issues. IEEE Commun Surv Tuts 25(4):2190–2244
11. Niu Y, Wei Z, Wang L, Wu H, Feng Z (2024) Interference management for integrated sensing and communication systems: a survey. arXiv:2403.16189
12. Cui Y, Koivunen V, Jing X (2018) Interference alignment based spectrum sharing for mimo radar and communication systems. In: Proceedings of 2018 IEEE 19th international workshop signal processing advanced wireless communication (SPAWC), pp 1–5
13. Bică M, Koivunen V (2019) Radar waveform optimization for target parameter estimation in cooperative radar-communications systems. IEEE Trans Aerosp Electron Syst 55(5):2314–2326
14. Ma L, Pan C, Wang Q, Lou M, Wang Y, Jiang T (2022) A downlink pilot based signal processing method for integrated sensing and communication towards 6G. In: Proceedings of 2022 IEEE 95th vehicular technology conference: (VTC2022-Spring), pp 1–5
15. Huusari T, Choi Y-S, Liikkanen P, Korpi D, Talwar S, Valkama M (2015) Wideband self-adaptive RF cancellation circuit for full-duplex radio: operating principle and measurements. In: Proceedings of 2015 IEEE 81st vehicular technology conference (VTC Spring), pp 1–7
16. Cao Y, Zhou J (2020) Integrated self-adaptive and power-scalable wideband interference cancellation for full-duplex MIMO wireless. IEEE J Solid-State Circuits 55(11):2984–2996

Chapter 8
Future Directions for FD-ISAC

Abstract This chapter discusses the potential future development directions of FD-ISAC technology. On one hand, the integration of FD-ISAC with other technologies, such as terahertz technology, RIS technology, and AI technology, offers greater performance gains for future communication and sensing. On the other hand, the future trend of FD-ISAC toward chip integration and miniaturization lays a foundation for its development in fields such as UAVs, and SAGIN. In addition to the challenges mentioned above, the following key areas also warrant further research and development, as detailed below.

8.1 Terahertz ISAC

Owing to the ever-increasing demand for spectrum, future wireless systems are migrating into higher-frequency bands—most notably millimeter-wave (30–100 GHz) and terahertz (0.1–10 THz)—to unlock vastly greater bandwidths than those available below 6 GHz. Millimeter-wave frequencies promise multi-gigahertz channels for next-generation networks [1, 2], while the terahertz realm not only affords exceptionally wide bandwidths for massive data throughput but also benefits from tiny wavelengths that enable compact hardware, ultra-low latency, and extremely high peak rates [3, 4].

Integrating full-duplex ISAC with these bands could simultaneously boost communications capacity and sensing fidelity. Yet managing self-interference (SI) becomes particularly challenging at millimeter-wave and terahertz frequencies, where massive antenna arrays and wide instantaneous bandwidths exacerbate leakage between transmit and receive paths. Conventional SIC methods designed for sub-6 GHz systems do not translate directly. To improve TX/RX isolation in millimeter-wave FD links, several beamforming-based solutions have been explored: purely analog schemes using networks of phase shifters [5–7], fully digital precoding with per-antenna RF chains [8], and hybrid analog-digital architectures that combine baseband beamformers with analog sub-arrays [9–11].

At terahertz frequencies, the same narrow beams that support high-precision communications also enable detailed environmental sensing, which can, in turn, refine

C. Du et al., *Full-Duplex Integrated Sensing and Communication Systems*,
https://doi.org/10.1007/978-981-92-0470-0_8

antenna alignment and enhance real-time localization and beam management [26]. Consequently, devising signal-processing algorithms tailored to ISAC operation in millimeter-wave and terahertz bands is a pivotal research direction.

Several hurdles remain. Practical millimeter-wave antennas exhibit imperfections—such as sidelobe leakage—that undermine isolation [12], mandating both RF-domain and digital-domain SIC stages. Moreover, much of the literature presumes SI is already suppressed (often via offline calibration [13, 14]), leaving the joint design of beamforming and power allocation for SI mitigation, capacity maximization, and sensing optimization largely unexplored. Finally, nearly all current FD-ISAC experiments focus on single-antenna setups; extending these concepts to massive arrays at millimeter-wave and terahertz frequencies represents an urgent and open challenge.

8.2 RIS Assisted ISAC

Reconfigurable intelligent surfaces (RIS)—also referred to as intelligent reflecting surfaces or large intelligent surfaces—have emerged as a transformative technology for sculpting the wireless propagation environment. Composed of a dense array of passive metamaterial elements whose reflection amplitudes and phases can be independently programmed via a central controller, RIS enables the constructive combination of multipath signals at targeted receivers, yielding the equivalent of 30–40 dB of passive array gain without the need for active RF chains or power amplifiers [15]. This exceptionally low-power, low-cost profile has made RIS a leading candidate for energy- and spectrum-efficient deployments in next-generation 6G networks [16–18].

When integrated into an ISAC framework, RIS brings the additional capability of actively tailoring the sensing environment. By judiciously configuring the element-wise phase shifts, an RIS can establish virtual line-of-sight paths or introduce new reflection angles to illuminate hidden regions, thereby extending detection coverage and improving target distinguishability in non-LoS scenarios [19]. Moreover, the large, reconfigurable aperture formed by the surface can be exploited to refine spatial resolution and enhance positioning accuracy, while simultaneously enabling dynamic interference suppression and communication-link enhancement through optimized reflection-coefficient matrices [20, 21].

Crucially, RIS itself operates as an inherently full-duplex reflector: it reradiates incident signals without generating self-interference or amplifying receive-path noise, in stark contrast to active multi-antenna transceivers. This property has inspired a growing body of work on RIS-assisted FD-ISAC systems, in which a full-duplex base station jointly optimizes its transmit beamformers, user transmit powers, and RIS phase profiles to balance communication throughput against radar-sensing fidelity. Recent studies have developed algorithms for maximizing sum-rate under radar SINR constraints [22], iteratively solving the coupled beamforming and power-allocation problems [23], and quantifying the fundamental trade-off between sensing and communication performance in this novel paradigm [24, 25]. Together, these advances underscore the promise of RIS-enabled FD-ISAC as a cost-effective, energy-efficient pathway toward simultaneous, high-accuracy sensing and high-capacity communications in future wireless networks.

8.3 AI-Driven ISAC

Artificial intelligence (AI) is poised to become a key enabler of next-generation 6G wireless networks, thanks to its ability to solve challenging non-convex optimization problems, to approximate complex system behaviors, and to uncover spatiotemporal patterns from large volumes of data. In particular, data-driven AI algorithms are creating new avenues for enhancing both FD and ISAC. By training deep or reinforcement learning models to map raw inputs—such as environmental observations, system configurations, and performance targets—directly to optimized transceiver settings, AI can substantially reduce reliance on conventional, computation-intensive channel estimation and modeling techniques, thereby improving resource allocation and overall system performance [26–28].

Within FD communications, AI-based methods offer promising alternatives for SIC. Traditional polynomial models for non-linear SI can become prohibitive in complexity and parameter count as the model order grows. Neural network-based cancellers, however, can achieve similar or better suppression with fewer parameters and floating-point operations [29, 30]. For example, architectures such as DN-lHLNN have been shown to reduce model size and computational load compared to classic polynomial approaches, while other designs—like channel-robust SI cancellers for time-varying channels [31], convolutional networks such as KD-MCNN [32], or neural-accelerated tuning of multi-tap RF cancellers [33]—demonstrate the versatility and efficacy of AI in tackling SIC challenges.

AI also brings significant advantages to ISAC, especially in scenarios plagued by poor channel conditions (e.g., indoor or densely built environments) that generate large amounts of noisy, multi-modal data with complex non-linear relationships. Conventional signal processing may struggle to disentangle communication and sensing tasks under such conditions [34]. Learned models—ranging from convolutional long short-term memory networks (HCL-Net) for predictive beamforming that balance radar accuracy with communication throughput [35], to unsupervised deep models that jointly optimize sensing and communications [36], and extreme learning machines for channel estimation in RIS-assisted ISAC systems [37]—offer robust, adaptable solutions that outperform traditional benchmarks.

Despite the aforementioned advancements, research on AI-driven FD-ISAC systems remains unexplored. For example, novel NN architectures and training methods are required to better address the challenges of separating target echoes from SI, to further improve SIC performance while avoiding the unintended removal of target echo signals in FD-ISAC system. Moreover, AI solutions have demonstrated the potential to reduce computational complexity for either SIC in FD radios or beamforming design in ISAC systems. However, how to leverage AI approaches to reduce the computational complexity of both remains an unresolved challenge. Furthermore, combining traditional mathematical methods with data-driven AI approaches has become a highly interesting and promising topic in FD-ISAC system, since it can effectively simplify the training process. Therefore, further exploration is needed for future advancements in AI-driven FD-ISAC systems.

8.4 ISAC for UAVs

Ground-based ISAC deployments often face significant performance degradation when line-of-sight (LoS) links are obstructed, and in urgent or unpredictable scenarios—such as disaster relief or post-attack response—establishing reliable terrestrial communication and sensing infrastructure may be infeasible. To mitigate these limitations, autonomous UAVs have been proposed as airborne ISAC platforms, leveraging their high mobility, unobstructed LoS from altitude, broad coverage potential, and rapid deployability [38]. Equipped with communication transceivers, UAVs can offer adaptable services to ground users when conventional networks are overloaded or disrupted.

In the UAV-assisted ISAC paradigm, aerial vehicles simultaneously serve as communication relays and sensing nodes, greatly enhancing LoS probability to both users and targets by virtue of their elevation [39]. This dual-function approach not only boosts overall system performance—potentially lowering power requirements compared to distributed ground systems and extending sensing range—but also reduces coverage gaps caused by terrain or man-made barriers. Moreover, the environmental data collected through sensing can inform more effective network topologies and refine wireless coverage strategies. By unifying communication and sensing on a single payload, UAV-ISAC systems eliminate the need for distinct devices, thereby decreasing both payload weight and energy consumption and extending mission endurance.

Recent studies have validated the promise of this approach. One cooperative UAV-ISAC architecture demonstrated a 66.3% gain in sensing performance over standard UAV networks [40], while another work formulated a periodic sensing-communication scheme that jointly optimizes UAV trajectory, transmit precoding, and sensing timing—maximizing user throughput under beam-pattern constraints [41]. Further research has developed frameworks for co-optimizing UAV flight paths and beamforming strategies to balance sensing accuracy against communication capacity across both quasi-stationary and fully mobile scenarios [42].

Despite these advances, the design and optimization of UAV-assisted ISAC remain challenging (particularly when incorporating FD operation). High UAV mobility complicates trajectory planning, necessitating the joint consideration of SIC, CCI, power budgets, speed limits, and waypoint constraints. At the same time, the stringent size, weight, and power restrictions on UAV platforms call for SI cancellers and

ISAC signal processors that are both low-complexity and energy-efficient. Although the added maneuverability introduces valuable spatial and temporal degrees of freedom for performance optimization, it also exponentially increases problem complexity. Consequently, developing robust, scalable algorithms that can withstand rapid dynamics and manage the elevated optimization burden is an essential focus for future UAV-assisted ISAC research.

8.5 ISAC for Space-Air-Ground Integrated Networks

Ground-based networks, high-altitude platforms, and satellites each offer unique advantages yet suffer inherent limitations when operated in isolation—ground systems face range restrictions from propagation losses [26], while aerial sensors contend with broader beam footprints and greater signal attenuation over extended distances [43]. The Space-Air-Ground Integrated Network (SAGIN) paradigm seeks to overcome these weaknesses by unifying these segments into a cohesive framework capable of delivering ubiquitous connectivity and sensing services on a global scale [43, 44]. By equipping nodes across the space, air, and ground layers with dual communication and sensing capabilities, SAGIN-enabled ISAC architectures can dramatically expand both coverage and functionality, enabling innovations such as perception-assisted communication—where sensing insights enhance link reliability—and networked perception, which fuses data from distributed nodes to elevate sensing fidelity [44].

Realizing ISAC within the multi-tiered SAGIN landscape, however, presents formidable challenges. It demands the simultaneous orchestration of two integration dimensions: the functional coupling of communication and sensing—potentially leveraging advanced full-duplex (FD) techniques—and the structural synchronization of heterogeneous network infrastructures spanning terrestrial, aerial, and orbital domains [45]. These intertwined complexities complicate system modeling, theoretical performance analysis, and joint optimization efforts. Addressing them will require substantial theoretical investigations and the development of scalable, robust algorithms tailored to SAGIN's dynamic, high-dimensional environment [45].

8.6 Conclusion

This chapter systematically elaborates on the future development directions of FD-ISAC. From a technological evolution perspective, its deep integration with cutting-edge technologies such as terahertz communication, RIS, and AI-empowered dynamic resource scheduling algorithms will significantly enhance the system's spectrum resource utilization, achieve sub-millisecond end-to-end latency, and construct an intelligent sensing system with cognitive decision-making capabilities. From a hardware implementation perspective, technological breakthroughs in heterogeneous integrated chip-level solutions and miniaturized reconfigurable antenna arrays will effectively promote the large-scale application of FD-ISAC in complex scenarios such as autonomous drone swarm collaboration and integrated SAGIN networking. It is worth emphasizing that subsequent research should focus on constructing a collaborative design paradigm that considers both theoretical innovation and engineering practice constraints, and prioritize breakthroughs in key technical bottlenecks such as multi-physics domain coupling modeling and cross-dimensional resource joint optimization, to provide systematic solutions for the industrialization of FD-ISAC technology.

References

1. Rappaport TS, Sun S, Mayzus R, Zhao H, Azar Y, Wang K, Wong GN, Schulz JK, Samimi M, Gutierrez F (2013) Millimeter wave mobile communications for 5G cellular: it will work! IEEE Access 1:335–349
2. Ayach OE, Rajagopal S, Abu-Surra S, Pi Z, Heath RW (2014) Spatially sparse precoding in millimeter wave MIMO systems. IEEE Trans Wireless Commun 13(3):1499–1513
3. Elbir A M, Mishra K V, Chatzinotas S, Bennis M (2022) Terahertz-band integrated sensing and communications: Challenges and opportunities. arXiv:2208.01235
4. Han C, Wu Y, Chen Z, Chen Y, Wang G (2024) THz ISAC: a physical-layer perspective of terahertz integrated sensing and communication. IEEE Commun Mag 62(2):102–108
5. Xiao Z, Chen S, Zeng Y (2024) Simultaneous multi-beam sweeping for mmWave massive MIMO integrated sensing and communication. IEEE Trans Veh Technol 73(6):8141–8152
6. Zhu L, Zhang J, Xiao Z, Cao X, Xia X-G, Schober R (2020) Millimeter-wave full-duplex UAV relay: joint positioning, beamforming, and power control. IEEE J Sel Areas Commun 38(9):2057–2073
7. Roberts IP, Vishwanath S, Andrews JG (2023) LoneSTAR: analog beamforming codebooks for full-duplex millimeter wave systems. IEEE Trans Wirel Commun 22(9):5754–5769
8. Liu X, Huang T, Shlezinger N, Liu Y, Zhou J, Eldar YC (2020) Joint transmit beamforming for multiuser MIMO communications and MIMO radar. IEEE Trans Signal Process 68:3929–3944
9. Roberts IP, Andrews JG, Vishwanath S (2021) Hybrid beamforming for millimeter wave full-duplex under limited receive dynamic range. IEEE Trans Wirel Commun 20(12):7758–7772
10. Zhao MM, Cai YL, Zhao MJ, Xu Y, Hanzo L (2020) Robust joint hybrid analog-digital transceiver design for full-duplex mmWave multicell systems. IEEE Trans Commun 68(8):4788–4802
11. Cai Y, Xu K, Liu A, Zhao M, Champagne B, Hanzo L (2020) Two-timescale hybrid analog-digital beamforming for mmWave full-duplex MIMO multiple-relay aided systems. IEEE J Sel Areas Commun 38(9):2086–2103
12. Lee H, Roberts IP, Lee N (2024) Nonlinear digital self-interference cancellation for side-lobe leakage in MIMO full-duplex systems. In: GLOBECOM 2024 - 2024 IEEE global communications conference. IEEE, pp 909–914
13. He Z, Xu W, Shen H, Ng DWK, Eldar YC, You X (2023) Full-duplex communication for ISAC: joint beamforming and power optimization. IEEE J Sel Areas Commun 41(9):2920–2936
14. Wang X, Fei Z, Zhang JA, Huang J (2022) Sensing-assisted secure uplink communications with full-duplex base station. IEEE Commun Lett 26(2):249–253
15. Poulakis M (2022) 6G's metamaterials solution: there's plenty of bandwidth available if we use reconfigurable intelligent surfaces. IEEE Spectr 59(11):40–45
16. Huang C, Zappone A, Alexandropoulos GC, Debbah M, Yuen C (2019) Reconfigurable intelligent surfaces for energy efficiency in wireless communication. IEEE Trans Wireless Commun 18(8):4157–4170

17. Liu Y, Liu X, Mu X, Hou T, Xu J, Di Renzo M, Al-Dhahir N (2021) Reconfigurable intelligent surfaces: principles and opportunities. IEEE Commun Surv Tuts 23(3):1546–1577
18. Verma S, Kawamoto Y, Kato N, Saiwai T, Yonehara M (2024) An efficient beam searching in hybrid intelligent reflecting/refracting surfaces (IRS)-aided mmWave 6G network. IEEE Trans Veh Technol 73(12):19299–19312
19. Hua M, Wu Q, Chen W, Fei Z, So HC, Yuen C (2024) Intelligent reflecting surface-assisted localization: performance analysis and algorithm design. IEEE Wireless Commun Lett 13(1):84–88
20. Li Z, Yuan W, Jing Z, Mu J, Wu N, Lin Z (2023) RIS-aided OTFS: a novel ISAC scheme for achieving real-time communication and sensing
21. Pogaku AC, Do DT, Lee BM, Nguyen ND (2022) UAV-assisted RIS for future wireless communications: a survey on optimization and performance analysis. IEEE Access 10:16320–16336
22. Zhang S, Hao W, Sun G, Huang C, Zhu Z, Li X, Yuen C (2024) Joint beamforming optimization for active STAR-RIS-assisted ISAC systems. IEEE Trans Wireless Commun 23(11):15888–15902
23. Guo Y, Liu Y, Wu Q, Li X, Shi Q (2024) Joint beamforming and power allocation for RIS aided full-duplex integrated sensing and uplink communication system. IEEE Trans Wireless Commun 23(5):4627–4642
24. Zhao X, Liu H, Gong S, Ju X, Xing C, Zhao N (2024) Dual-functional MIMO beamforming optimization for RIS-aided integrated sensing and communication. IEEE Trans Commun 72(9):5411–5427
25. Le QN, Nguyen VD, Dobre OA, Shin H (2023) RIS-assisted full-duplex integrated sensing and communication. IEEE Wirel Commun Lett 12(10):1677–1681
26. Gan T, Dang S, Li X, Zhang Z (2024) Integrated sensing and communications for 6G: prospects and challenges of using THz radios. In: 2024 IEEE Wireless Communications and Networking Conference (WCNC). IEEE, pp 1–6
27. Liu X, Zhang H, Sun K, Long K, Karagiannidis GK (2024) AI-driven integration of sensing and communication in the 6G Era. IEEE Netw 38(3):210–217
28. Wu N, Jiang R, Wang X, Yang L, Zhang K, Yi W, Nallanathan A (2024) AI-enhanced integrated sensing and communications: advancements, challenges, and prospects. IEEE Commun Mag 62(9):144–150
29. Elsayed M, El-Banna AA, Dobre OA, Shiu W, Wang P (2022) Full-duplex self-interference cancellation using dual-neurons neural networks. IEEE Commun Lett 26(3):557–561
30. Elsayed M, El-Banna AAA, Dobre OA, Shiu WY, Wang P (2024) Machine learning-based self-interference cancellation for full-duplex radio: approaches, open challenges, and future research directions. IEEE Open J Veh Technol 5:21–47
31. Kong DH, Kil YS, Kim SH (2022) Neural network aided digital self-interference cancellation for full-duplex communication over time-varying channels. IEEE Trans Veh Technol 71(6):6201–6213
32. Wang X, Zhao H, He Y, Hu P, Shao S (2024) A simple neural network for nonlinear self-interference cancellation in full-duplex radios. IEEE Trans Veh Technol 73(7):10817–10822
33. Kolodziej KE, Cookson AU, Perry BT (2021) RF canceller tuning acceleration using neural network machine learning for in-band full-duplex systems. IEEE Open J Commun Soc 2:1158–1170
34. Liu A, Huang Z, Li M, Wan Y, Li W, Han TX, Liu C, Du R, Tan DKP, Lu J, Shen Y, Colone F, Chetty K (2022) A survey on fundamental limits of integrated sensing and communication. IEEE Commun Surv Tuts 24(2):994–1034
35. Liu C, Yuan W, Li S, Liu X, Li H, Ng DWK, Li Y (2022) Learning-based predictive beamforming for integrated sensing and communication in vehicular networks. IEEE J Sel Areas Commun 40(8):2317–2334
36. Qi Q, Chen X, Zhong C, Yuen C, Zhang Z (2024) Deep learning-based design of uplink integrated sensing and communication. IEEE Trans Wireless Commun 23(9):10639–10652

37. Liu Y, Al-Nahhal I, Dobre OA, Wang F, Shin H (2023) Extreme learning machine-based channel estimation in IRS-assisted multi-user ISAC system. IEEE Trans Commun 71(12):6993–7007
38. Li B, Fei Z, Zhang Y (2019) UAV communications for 5G and beyond: recent advances and future trends. IEEE Internet Things J 6(2):2241–2263
39. Liu F, Cui Y, Masouros C, Xu J, Han TX, Eldar YC, Buzzi S (2022) Integrated sensing and communications: toward dual-functional wireless networks for 6G and beyond. IEEE J Sel Areas Commun 40(6):1728–1767
40. Chen X, Feng Z, Wei Z, Gao F, Yuan X (2020) Performance of joint sensing-communication cooperative sensing UAV network. IEEE Trans Veh Technol 69(12):15545–15556
41. Meng K, Wu Q, Ma S, Chen W, Quek TQS (2022) UAV trajectory and beamforming optimization for integrated periodic sensing and communication. IEEE Wireless Commun Lett 11(6):1211–1215
42. Lyu Z, Zhu G, Xu J (2023) Joint maneuver and beamforming design for UAV-enabled integrated sensing and communication. IEEE Trans Wireless Commun 22(4):2424–2440
43. Zhang Y, Wang J, Li Q, Chen J, Feng H, He S (2024) Joint communication, sensing, and computing in space–air–ground integrated networks: system architecture and handover procedure. IEEE Trans Mobile Comput 19(2):70–78
44. Cui H, Zhang J, Geng Y, Xiao Z, Sun T, Zhang N, Liu J, Wu Q, Cao X (2022) Space-air-ground integrated network (SAGIN) for 6G: requirements, architecture and challenges. China Commun 19(2):90–108
45. Mao W, Lu Y, Pan G, Ai B (2025) UAV-assisted communications in SAGIN-ISAC: mobile user tracking and robust beamforming. IEEE J Sel Areas Commun 43(1):186–200

Index

C. Du et al., *Full-Duplex Integrated Sensing and Communication Systems*,
https://doi.org/10.1007/978-981-92-0470-0

GPSR Compliance

The European Union's (EU) General Product Safety Regulation (GPSR) is a set of rules that requires consumer products to be safe and our obligations to ensure this.

If you have any concerns about our products, you can contact us on ProductSafety@springernature.com

In case Publisher is established outside the EU, the EU authorized representative is:

Springer Nature Customer Service Center GmbH
Europaplatz 3
69115 Heidelberg, Germany

Batch number: 10406147

Printed by Printforce, the Netherlands